6-19-74

FARMER DISCONTENT, 1865-1900

PROBLEMS IN AMERICAN HISTORY

EDITOR
LOREN BARITZ
State University of New York, Albany

Ray B. Browne POPULAR CULTURE AND THE EXPANDING
CONSCIOUSNESS
Vernon R. Carstensen Farmer Discontent, 1865-1900
William N. Chambers THE FIRST PARTY SYSTEM:
Federalists and Republicans
Don E. Fehrenbacher THE LEADERSHIP OF ABRAHAM
LINCOLN
Eugene D. Genovese THE SLAVE ECONOMIES
Volume I: Historical and Theoretical Perspectives
Volume II: Slavery in the International Economy
Paul Goodman THE AMERICAN CONSTITUTION
Richard J. Hooker THE AMERICAN REVOLUTION
Richard Kirkendall THE NEW DEAL:
The Historical Debate
Stephen G. Kurtz THE FEDERALISTS—
Creators and Critics of the Union 1780–1801
Walter LeFeber AMERICA IN THE COLD WAR
Walter LeFeber THE ORIGINS OF THE COLD WAR, 1941–1947
Donald G. Mathews AGITATION FOR FREEDOM:
The Abolitionist Movement
Douglas T. Miller THE NATURE OF JACKSONIAN AMERICA
Richard H. Miller AMERICAN IMPERIALISM IN 1898
Thomas J. Pressly RECONSTRUCTION
Richard Reinitz TENSIONS IN AMERICAN PURITANISM
Darrett B. Rutman THE GREAT AWAKENING
David F. Trask WORLD WAR I AT HOME
Patrick C. T. White THE CRITICAL YEARS, AMERICAN
FOREIGN POLICY, 1793–1825
Harold D. Woodman THE LEGACY OF THE AMERICAN ,
CIVIL WAR

FARMER DISCONTENT
1865-1900

EDITED BY

Vernon Carstensen

John Wiley & Sons, Inc.
New York • London • Sydney • Toronto

Library of Congress Cataloging in Publication Data:

Carstensen, Vernon Rosco, 1907—   comp.
      Farmer discontent, 1865–1900.

        (Problems in American history series)
        1. Agriculture and politics—United States—
History—Addresses, essays, lectures.   2. Patrons
of Husbandry—Addresses, essays, lectures.   3. People's
Party of the United States—Addresses, essays, lectures.
I. Title.

HD1759.C37   322′.3′0973   73-15951
ISBN 0-471-13724-3
ISBN 0-471-13725-1 (pbk.)

Printed in the United States of America

10 9 8 7 6 5 4 3 2 1

SERIES PREFACE

This series is an introduction to the most important problems in the writing and study of American history. Some of these problems have been the subject of debate and argument for a long time, although others only recently have been recognized as controversial. However, in every case, the student will find a vital topic, an understanding of which will deepen his knowledge of social change in America.

The scholars who introduce and edit the books in this series are teaching historians who have written history in the same general area as their individual books. Many of them are leading scholars in their fields, and all have done important work in the collective search for better historical understanding.

Because of the talent and the specialized knowledge of the individual editors, a rigid editorial format has not been imposed on them. For example, some of the editors believe that primary source material is necessary to their subjects. Some believe that their material should be arranged to show conflicting interpretations. Others have decided to use the selected materials as evidence for their own interpretations. The individual editors have been given the freedom to handle their books in the way that their own experience and knowledge indicate is best. The overall result is a series built up from the individual decisions of working scholars in the various fields, rather than one that conforms to a uniform editorial decision.

A common goal (rather than a shared technique) is the bridge of this series. There is always the desire to bring the reader as close to these problems as possible. One result of this objective is an emphasis of the nature and consequences of problems and events, with a de-emphasis of the more purely historiographical issues. The goal is to involve the student in the reality of crisis, the inevitability of ambiguity, and the excitement of finding a way through the historical maze.

Above all, this series is designed to show students how experienced historians read and reason. Although health is not contagious, intellectual engagement may be. If we show students something significant in a phrase or a passage that they otherwise may have missed, we will have accomplished part of our objective. When students see something that passed us by, then the process will have been made whole. This active and mutual involvement of editor and reader with a significant human problem will rescue the study of history from the smell and feel of dust.

Loren Baritz

CONTENTS

Part One   GROWTH AND CHANGE IN AMERICAN
AGRICULTURE, 1865-1900   1

Part Two THE GRANGER MOVEMENT, 1865-1880   17

1. *A Farmers' Platform* Adopted at Centralia, Illinois,
September 15, 1858   19
2. Jonathan Periam, *Illinois Farmers Demand State
Control of Railroad, 1869*   21
3. Dudley W. Adams, *A Farmer Speaks About Farmers
and Farming, 1872*   26
4. *The Constitution of the National Grange of the
Patrons of Husbandry, 1873*   32
5. *The Illinois Farmers' Association, 1873*   35
6. *The Declaration of Purposes of the Patrons of Husbandry, 1874*   40
7. Willard G. Flagg, *An Illinois Farmer Talks About
the Farmers' Movement, 1874*   45
8. *Charles Francis Adams Offers Judgement on the
Granger Movement, 1875*   54
9. *The Nation Speaks About the Grangers, 1876*   65

Part Three WHEELS, ALLIANCES, AND POPULISM   69

10. W. Scott Morgan, *The Changing Purposes of the
Farmers' Alliances*   71
11. James Baird Weaver, *The Threefold Contention of
Industry*   80
12. K. H. Porter and D. B. Johnson, *The People's Platform, 1892*   89
13. Frank Basil Tracy, *Rise and Doom of the Populist
Party*   94
14. Charles S. Gleed, *The True Significance of Western
Unrest*   101

15. K. H. Porter and D. B. Johnson, *Party Platforms, 1896*   107
16. Henry Demarest Lloyd, *The Populists at St. Louis*   124
17. Newell D. Hillis, *An Outlook upon the Agrarian Propaganda in the West*   134
18. Edwin Markham, *The Man with the Hoe*   139
19. William V. Allen, *The Populist Program, 1900*   141

Part Four   THE VIEW FROM THE STUDY   145

20. John D. Hicks, *The Legacy of Populism in the Western Middle West*   149
21. C. Vann Woodward, *The Populist Heritage and the Intellectual*   155
22. Oscar Handlin, *Reconsidering the Populist*   173

Part Five   Bibliographical Note   181

PART ONE

Growth and Change in American Agriculture, 1865 to 1900

The years from the Civil War to 1900 were marked by unprecedented expansion in the number of farms in the United States, impressive technological advances in farm machinery, the rise of agricultural education, a substantial movement to bring science to the aid of the farmer, and a spectacular increase in farm production. This was also a time of the organization and spread of farmer protest movements—the "Granger Movement," the "Agrarian Crusade," the "Populist Revolt"—all of which heralded widespread discontent among large numbers of farmers. During these years agriculture prospered wonderfully but many farmers did not.

James Bryce declared in *The American Commonwealth,* published in 1888, that western America (he did not specify precisely what he included in this term) was "one of the most interesting subjects of study the modern world has seen. There has been nothing in the past resembling its growth, and probably there will be nothing in the future." This vast region, he continued, was possessed of a good climate, abundant and varied resources in fertile soil, rich mineral deposits and vast forest. ". . . the whole of this virtually unoccupied territory [was] thrown open to an energetic race, with all the appliances and contrivances of modern science at its command—these are phenomena absolutely

1

without precedent in history, which cannot recur elsewhere, because our planet contains no such other favoured tract of country." [James Bryce, *The American Commonwealth* (New York,
1888. Three volumes) , *3*:634.]

During the four decades between 1860 and 1900 the total
population of the United States had increased from 31.5 to 76
million despite the destruction and disruption of the Civil War
and Reconstruction. In 1860 California and Oregon were the
only states organized in the vast region west of the Missouri River; by 1900 only Arizona, New Mexico and Oklahoma remained
as territories. Never in American history either before or since
would so many new farms be made in so short a time. In 1860
the federal census reported 2 million farms; in 1900 the number
had reached 5.7 million, an increase of 3.7 million. In 40 years
almost twice as many farms had been made as were made in all
the years between 1607 and 1860. The highest total number of
farms in the history of the Republic was reached in 1935 when
the census reported 6.8 million. Improved acreage on farms increased from 163 million in 1860 to 415 million in 1900. To the
extent that federal land policies governing the distribution of
land in the public domain were aimed at turning the public lands
into farms as quickly as possible, this great growth in farming reflects a magnificent success of these policies.

The large growth in the number of farms and in cultivated
acreage was accompanied by a striking increase in the total production of American farms. The increase is suggested in the census reports on the three great staple crops, cotton, wheat, and
corn. Once in the 1850s cotton production approached 5 million
bales, but mostly it ranged between 2 and 3 million. Before the
end of the century, cotton production had passed the 10 million
mark. Wheat production increased from 173 million bushels in
1860 to 600 million in 1900, corn from 838 million bushels to
2600 million. The director of the census declared that, in 1900,
agricultural exports alone "now have an annual value nearly, if
not quite, equal to one half of the total production of staples in
1850."

The numbers that show the total increase in numbers of farms
and bales, bushels, or pounds of farm produce and livestock do
show the total production for the nation, but they fail to differen-

American Settlement 1860–1900

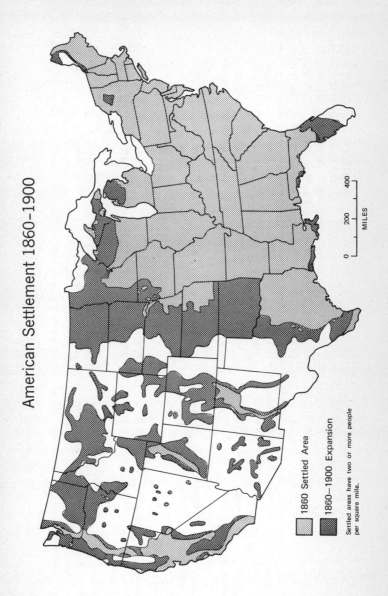

1860 Settled Area

1860–1900 Expansion

Settled areas have two or more people per square mile.

0   200   400

MILES

Table 1 Growth of Agriculture 1860 to 1900

		North Atlantic (Conn., Maine, Mass., N. H., N. J., N. Y., Pa., R. I., Vt.)	South Atlantic (Del., D. C., Fla., Ga., Md., N. C., S. C., Va., W. V.)	North Central (Ill., Ind., Ia., Kans., Mich., Minn., Neb., Mo., O., N. D., S. D., Wis.)	South Central (Ala., Ark., Ky., Indian Territory, La., Miss., Okla., Tenn., Tex.)	West (Ariz., Calif., Col., Ida., Mon., Nev., N. M., Ore., Wash., Wyo., Utah.)	Total
Number of farms	1860	564,935	301,165	772,165	370,373	34,664	2,044,077
	1900	677,506	962,225	2,196,567	1,658,166	242,908	5,789,657
Total value (to nearest million)	1860	$2,454	$1,207	$2,523	$1,672	$123	$7,980
	1900	2,951	1,454	11,505	2,816	1,715	20,514[a]
Total value of farm products (to nearest million)	1870[b,c]	$634	$308	$979	$457	$70	$2,448
	1900	666	465	2,360	889	337	4,739
Total number of horses and mules	1860	1,297,182	1,014,543	2,706,782	2,114,662	234,422	7,400,322
	1900	1,687,428	1,560,383	9,722,726	4,853,476	2,161,165	20,004,223
Amerage size of farms in acres	1860	108.1	352.8	139.7	321.3	366.9	199.2
	1900	96.5	108.4	144.5	155.4	386.1	146.6

SOURCE. Census Reports, *Twelth Census of the United States Taken in 1900, Vol. 5, Agriculture. Pt. I. Farms Livestock, and Animal Products*, GPO 1902, p. xvi, ff.
[a] The five divisions plus Alaska and Hawaii.
[b] Not reported in 1860.
[c] Values in 1870 were reported in depreciated currency. To reduce to the specie basis of other figures they must be diminished one fifth.

tiate between the several regions of the country. Prosperity and
increased production were not distributed evenly across the
country, nor even among the farms within the different regions
into which the managers of the census divided the country. The
map (Figure 1) and the chart (Table 1) suggest the vastly different rates of growth in the several regions and the differences
in productivity.

The most extensive growth in the number of farms had occurred in the North Central, South Central and Western regions.
The large increase in the number of farms in the South Atlantic
region reflects some new land settlement but can be accounted
for largely by a change from the large plantation system to a system of sharecropping. The North Central region by 1900 accounted for almost half the dollar value of farm production and
also claimed almost half the total value of all farms in the nation.

There were substantial differences between the various regions as established by the census, but there was also a vast difference across the country between the productive units counted
as farms. The term *farm* was sufficiently elastic to embrace a
rock-strewn hillside farm in Vermont about to be abandoned, a
three- or four-acre cotton patch held by an Arkansas sharecropper, a full section (640 acres) corn-hog farm in the fat rich land
of north-central Illinois, a miserable stump farm in the Cutover
of Northern Wisconsin with its only building a tarpaper-covered
shack, a bonanza wheat farm in the Red River Valley that in the
1890s employed steam engines for plowing and for threshing, a
raw, hardscrabble homestead in western Kansas or Nebraska
with its sod house and pole and straw shed for livestock, a prosperous citrus ranch in California, the XIT ranch in Texas, a
one-acre truck farm in New Jersey, and a greenhouse or kosher
dairy on Manhattan Island. But whether large or small, well-
managed or neglected, massively productive or a dried out piece
of land that barely made a crop, each was counted a farm and
together they provided the data that census officials and others
used to report on the character and dimensions of American agriculture.

In 1860 the census reported that of the 2 million farms counted that year, 822,697 were farms of under 50 acres and another
607,688 ranged from 50 to 100 acres. In 1900, of the total of
5.7 million farms, only 2.4 million included more than 100

acres. It is impossible to generalize about these data because some farmers on small holdings of rich land near good markets might prosper; others, with large holdings of productive or unproductive land far from markets never tasted prosperity. These data do suggest, however, the wide variety in the holdings that made up a farm and should warn one against making or accepting easy generalizations about farms and farmers.

The tools, the machines, and the energy used in farming changed extensively during this period of rapid growth. In 1860 men, women, children, and draft animals (work oxen, horses, mules and a few donkeys) did most of the work needed for farming whether it was raising cotton, tobacco, corn, wheat, or other cereals, tending livestock, truck gardening, or fruit raising. Work oxen were still used in large numbers, but horses and mules were becoming more widely used. New farm machines, reapers, headers, mowers, the first horse-drawn corn planters, hayracks, hayforks, and other machines recently invented, required a faster step than that of the slow-footed oxen. In 1850, the first agricultural census reported 1.7 million work oxen, 4.3 million horses, and one-half million mules and asses. By 1860 work oxen numbered 2.3 million, horses 6.2 million, and mules and asses 1.1 million. By 1900 the number of horses in the nation (not all on farms) had increased to 21.4 million, mules and asses to 3.4, and work oxen were no longer counted. When last reported, in 1890, the number had diminished to 1.1 million. In 1900 they were still used on backward farms but were not counted separately from other cattle.

The vast agricultural growth of the North Central Region was probably foreshadowed by the fact that Cyrus Hall McCormick moved his reaper manufacturing plant to Chicago in the late 1840s and J. I. Case had already begun to manufacture his remarkably successful threshing machine in Racine, Wisconsin. The 1860 census reported that there were almost 2000 farm implement manufacturers in the country, of which 840 were located in the western states, and that their plants produced one half of the $17.5 million dollars worth of farm machines manufactured that year.

By 1900 many of the promises of 1860 were being fulfilled. Horses and mules provided most of the power needed for cultivation and harvesting, but the steam engine was widely used to furnish power for threshing and some other farm work; and stationary internal combustion engines had appeared. The transformation of the wheat harvest had been dramatic. In 1860 the scythe and cradle, the threshing floor, and winnowing of the grain were still well known, even in areas where horse-drawn reapers and horse-powered threshing machines were used. By 1900 the reaper had been superceded by the binder, and in the large wheat-producing areas the header had been joined to a threshing machine to become the grain combine. Some of these machines had cutting bars from 12 to 20 feet in length and were drawn by from 20 to 30 or more horses. In California a few wheat farmers boasted of gigantic machines drawn by steam tractors and equipped with cutting bars of 24 to 40 feet in length. It was claimed that such machines could cut from 60 to 120 acres of grain in a full working day. In the 1840s, when most grain was still cut by men with scythe and cradle, it was assumed that a good man could cut one or two acres of grain in a 10- to 12-hour day, and there still remained the work of tying the cut grain into bundles and shocking the bundles for drying. When the grain was dry, the bundles were carried to a central point, built into a stack, threshed, and the grain then cleaned by winnowing it or putting it through a fanning mill propelled by men or animals.

Not all farming benefited equally from the advances in machinery. Producers of cotton and tobacco shared much less in the improvement and invention of tools and machines than did producers of wheat and corn.

The striking advances in farm machinery and the vast expansion in the number of farms and acreage under cultivation both fed, and fed on the equally spectacular changes in other parts of the nation's economy. By 1860, 31,000 miles of railroad had added another dimension to the emerging transportation network that included roads, steamboats, and other craft on the rivers and the lakes, and the newly built canals that connected the great Lakes to the Hudson River and the eastern seaboard, and the lakes to the Ohio-Mississippi rivers. From the first farm settlement in the Ohio Valley, farm produce sought a market mostly along the water routes of the Ohio and the Mississippi rivers and to the great markets of the world. The firs. decade of the 19th century had also seen herds of corn-fattened ca. marched eastward from Ohio to the seaboard markets. After l. 1840s railroads began to carry the cattle east. In the 1850s the first railroad bridges had been thrown across the Mississippi.

By 1900 the census reported close to 200,000 miles of railroad with a half-dozen lines reaching the Pacific coast. Whe. from the Red river valley of the North, and from Kansas, hundreds of miles from good water routes, now sought and reached a European market; California oranges, Oregon apples and pears, and other western produce could seek and claim markets to the east and abroad.

The mechanisms and the organizations needed to gather, store, process, and market farm produce grew enormously, and new ones were devised in the wake of the expanding transportation system. Grain elevators, warehouses, and huge stockyards came into existence along with refrigerator cars for railroads and refrigerator ships which permitted the export, beginning in the 1870s, of what one Englishman disdainfully called dead beef to the English and other foreign markets. Rollers supplanted the slow-moving millstones in the flouring mills. By the 1880s food from the American farms had become so cheap and abundant it was asserted that an English workman could afford to eat white bread and beef. This was good or bad depending on the point of view. Each of the new or enlarged agencies that touched farm produce on its way from farmer to consumer levied its tribute for "service performed," and the farmer got what was left after all these costs were met, or so it seemed to him. All of these agencies came to be embraced in one word, middleman.

The enlarged transportation system also invited those who produced tools and machines and other things the farmer needed —tools and machines he had once made for himself or had had made by a local mechanic or blacksmith—to seek a regional or national market. The McCormick and Case companies illustrate this development. Even before the Civil War, McCormick was supplying reapers to California farmers and to farmers in Europe. Similarly, the large meat-packing and other food-processing industries represent the concentration and industrialization of functions once performed in whole or in part on the farm or in the service villages of the farming areas.

Thus in the years that the number of farmers and farms increased most rapidly, the farmers' place in the vast economic scheme of things changed greatly and was greatly diminished. During this period the farm labor force had grown from 6.2 to 10.9 million; the nonfarm labor force had grown from 4.3 to 18.1 million.

The rise of the agricultural colleges in the years after the Civil War represented another aspect of change along with the steps taken to bring science to the aid of the farmer and the creation of the United States Department of Agriculture. All of these developments were deeply rooted in the American past, but the period after the Civil War was the time they emerged conspicuously.

As early as the 1780s, public-spirited men in Pennsylvania, South Carolina, and other states, often seeking to imitate similar actions in England and on the Continent, formed agricultural societies dedicated to the improvement of agriculture in all its aspects. The Massachusetts Agricultural Society, of which John Adams once served as president, offered a reward to "men of enterprise who have, by their inquiries, made useful discoveries and communicated them to the public." When Thomas Jefferson wrote the constitution of the Albermarle Agricultural Society, he included a provision calling for reports on all the productive and unproductive farming practices in the region in the expectation that "the choicest processes culled from every farm would compose a course probably near perfection." Such sentiments are found reflected again and again in the statement of purposes adopted by the farmers clubs that appeared in many farm communities created in the westward march of settlers. These clubs which met periodically sought, by pooling their knowledge and their experience in the new lands they occupied, to improve the art and craft of the farmer members. This committment to a more productive, more efficient agriculture was reflected in other ways: in the organization of agricultural fairs, in the support, sometimes niggardly, of farm journals, in the collection of agricultural statistics, and in the creation of agricultural colleges.

Proposals for agricultural colleges had been made before the end of the 18th century. By 1860 provisions had been made in Maryland, Pennsylvania, Michigan, and Iowa for institutions of this kind to be created and supported at state expense. In 1862 Congress adopted the first Morrill Act, which provided a grant of public land to each state not then in rebellion against the central government to be used for the establishment of an agricultural college. After the war the act was extended to all states.

There was much difficulty initially in determining what the new colleges would be and do, but in the years after 1862 a teaching staff was found, a course of study gradually took shape, and leaders of the colleges formed a national organization for the exchange of ideas and for mutual support.

The spirit that called the colleges of agriculture into existence also accounted for the creation of the United States Department of Agriculture in 1862. President Abraham Lincoln in his first annual message reminded Congress that the important interest of agriculture was represented in the federal government by no department, no bureau, "but a clerkship only," and suggested that something more be done. George Washington in his last message to Congress had urged that an agricultural board be created that would collect, organize, and disseminate agricultural information "to encourage and assist a spirit of discovery and improvement," but Congress then had been deaf to the proposal. To Lincoln's suggestion Congress did respond with a law creating a department of agriculture which among other things, was to "acquire and diffuse among the people of the United States useful information on subjects connected with agriculture in the most general and comprehensive sense of the word." It was to collect, test, and distribute new and valuable plants and seeds and to conduct "practical and scientific experiments." The committee that recommended the creation of the department declared that ". . . the man who makes two blades of grass grow where one grew before is the benefactor of his race."

The new department grew and exfoliated in the years that followed. It employed scientists who sought to aid farmers by developing better plants, in searching the world for new species of useful farm plants, and in finding ways to control plant and animal diseases. In 1889 the commissioner of the department was made a member of the President's cabinet, partly in response to demands of farmer organizations.

All the enginery of society—the economic, the educational, and the other social institutions—seemed, in these years after the Civil War, to encourage more land settlement and more efficient farming and thus to push the farmers and farming more and more toward complete enfoldment in the commercial system just as at an earlier time the tobacco and the cotton farmers had become one-crop producers dependent for their livelihood on the market and those who managed it. Yet ancient and romantic beliefs continued to picture farming as the idyllic, independent way of life. It was not a way of making a living, but a way of living and farmers were a fortunate, superior people. Thus Jefferson had written in his *Notes on Virginia* that "Those who labor in the earth are the chosen people of God, if ever He had a chosen people, whose breasts he has made His peculiar deposit for substantial and genuine virtue . . . The mobs of the great cities add just so much to the support of pure government as sores do to the strength of the human body". The farmer, wrote the *Union Agriculturalist and Western Prairie Farmer* in 1841, "is the most noble and independent man in society" (quoted in *Yearbook of Agriculture*, 1940, p. 117). "Burn down your cities," cried William Jennings Bryan in Chicago in July, 1896, "and leave our farms, and your cities will spring up again as if by magic; but destroy our farms and the grass will grow in the streets of every city in the country."

Such views of farmers and farming were, of course, denied by references to rubes, hicks, hayseeds, bumpkins, rustics, and clodhoppers, but even so, the 19th-century farmer in America was widely pictured as a self-subsistent, independent, free man who dwelt in peace under his own fig tree, his granary full and overflowing, his flocks and herds as numerous and thriving as those of an Old Testament patriarch, his family about him, well clothed, well sheltered, well fed, all untouched by the sordid lust for gain that marked the unhappy and unfortunate wretches condemned to endless and unrewarding toil in the teeming, sinful cities.

Such a view of farmers and farming was probably most fiercely held by editorial writers for urban newspapers, preachers, and politicians who had themselves never hoed cotton or corn, or shocked, stacked, and threshed bearded wheat or barley under a hot sun when the temperature stood at 90 degrees or higher, milked a dozen cows night and morning, worked 14 to 16 hours a day, six days a week, from spring till fall, or waded ankle deep through mud in the barnyard every time it rained or thawed throughout the year.

That the idyllic view of farmers was deeply and widely held is no doubt reflected in the public cry of outrage that spread across the land after the *San Francisco Examiner* published Edwin Markham's "Man With the Hoe" on January 15, 1899. The romantic view of farming and farmers held some elements of truth and it was useful to poets, preachers, and politicians, but for farmers adrift in a sea of change, it offered nothing that would help them to find their bearings and to set a course.

Farmers, an infinitely various group, were both the beneficiaries and the victims of the massive changes and the headlong expansion of agriculture. One certainty they could depend on, if they were willing to face it, was that nothing would remain the same for very long. Thus when new lands were settled west of the Appalachian mountains, wheat farmers in the eastern states discovered they had new competitors for their markets, and they must adjust to the new competition, just as those wheat farmers of Ohio in time would have to adjust to those still further west. In 1880 a Wisconsin farmer, in urging farmers there to change from raising wheat to dairying, declared: "It would be folly to close our eyes to the fact that we are now menaced with an almost unlimited area of rich, cheap land that can pour productions into our market far below the cost of productions consistent with the price of land in Wisconsin. . . ." He also warned that "Just now Wisconsin farmers have a protection in what we please to call railroad extortions, in freight from the far west," but this situation would not last.

Farmers had to adjust and accommodate quickly to the new technology and the new business relationships even if they were able to continue producing what they knew. Sometimes this was not possible within the range of productivity set by the land itself. Thus new farming areas in the west spelled ruin for many

New England farms, a melancholy fact written with increasing boldness in the record of farm abandonment in the years after the Civil War. The cotton sharecropper with his tiny patch of land, his ancient methods and his antiquated tools, probably held on in misery and poverty simply because there was no other place to go.

But the greater efficiency in farming, even for the successful farmers, did not offset the almost steady decline in farm prices. What took place is suggested in the prices of the great staples: cotton, wheat, and corn. These prices, gathered by the federal census, ~e probably higher than those actually received on farms.

The price of co ˑn stood at 16.5 cents a pound in 1869. Although the trend was ˑwn, the price did not drop below 10 cents until 1875. In the lo.°ˑ it ranged from 5 to 8 cents. A bumper crop of more than 10 m.ˑ·ˑn bales sent prices below 5 cents a pound in 1894. The whole cou. ˑron of a little over 4 million bales in 1870, with a price of 12 cen.. ˑ , ˑˑd. fetched more than the 10-million-bale crop in 1894. Wheat was ˑ ˑrted at over $2 a bushel in 1866; after 1881 it dropped below a aˑ. lar a bushel and stayed there until well into the 20th century. The whole crop of over 541 million bushels in 1894 fetched only slightly more than a 210-million-bushel crop had brought in 1876. Corn prices were reported at 65 cents a bushel in 1866, 78 cents the next year, fell thereafter to below 50 cents by 1871, and to 27 cents in 1889. In 1895 an enormous crop of 2.5 billion bushels, with an average price of a fraction over 25 cents, had only a slightly higher dollar value than a 794 million bushel crop in 1867. These figures show very bad years against good years but in 1867, for example, farmers harvested 32 million acres of corn; in 1895 they harvested 90 million acres for approximately the same return in dollars. The years that corn fell below 30 cents a bushel were years when farmers burned corn for fuel because it was cheaper than wood or coal.

Crude data such as that offered above says nothing about the disasters faced by individual farmers; it does suggest the broad dimensions of the problem of the farmer who participated in the enormous prosperity of agriculture but may have invited bankruptcy for his efforts.

But farmers were not only plagued by declining prices—their reward for laboring from sunrise to sunset—they were caught and held in what they viewed as an unfair scheme of things. The products of their labor were sold in a market that fluctuated according to the place and the amount of goods, but the prices a farmer had to pay for tools, machines, seed, fertilizer, transportation, storage in warehouses, elevators and stockyards, to commission merchants, for interest on crop loans and mortgages, and for taxes, were or seemed to be fixed and unresponsive to the downward movement of farm prices. Charges that could be borne without much complaint in a rising market appeared to be exorbitant when prices of farm goods fell disastrously. It was one thing to pay 20 cents a bushel to market wheat when the price stood at $2 a bushel; it was quite another when the price stood at 80 cents. Complainers hardly noticed that freight rates had been reduced. Added to the complaint of "exorbitant" prices was the greater and more emotional complaint that railroads and other middlemen discriminated between large companies and individuals, between regions, and in other ways.

Farmers responded in many ways to the situation in which they found themselves. Like the tobacco farmers of colonial times and the cotton farmers before the Civil War, they organized associations designed to eliminate the middleman and his profits. The Patrons of Husbandry and related organizations that came to be identified as the Granger Movement encouraged creation of cooperatives that would buy and sell for the farmers, and some even sought to manufacture farm machinery, process farm products, and build and manage grain elevators and warehouses. Most of these ventures failed, but some survived and served a useful purpose; they did not and probably could not attain the goals their supporters set. Many of the objectives sought, for instance, the control of railroad and warehouse and elevator charges, could be obtained or so the farmers and their allies claimed, only through legislation that would place these agencies under public control and fix their rates. Even though the Patrons of Husbandry, and the later farm Alliances asserted that they would eschew partisan politics, the programs they set for themselves required laws to curb and regulate the avarice and conduct of railroad, warehouse, elevator, and other companies. The

Granger Movement of the late 1860s and early 1870s sought to capture state legislatures to obtain laws to control the railroads, elevators, and warehouses, and they had some success in the midwestern states, notably Illinois, Iowa, Wisconsin, and Minnesota. The Granger Movement subsided in the middle and late 1870s, but in the 1880s the Alliance Movements rose in the wheat and cotton country. The aspirations of the Alliances spawned the Populists' Movement which, in its platforms of 1892 and 1896, sought specific and general reforms. Althou n victory was denied in the election of 1896, the Populists set goals for political and economic reform that would color the conduct of both major parties far into the 20th century.

PART TWO

The Granger Movement, 1865 to 1880

The Granger Movement is a term used by historians and others to embrace both the formation and growth of the first large national farmers organization, the National Grange of the Patrons of Husbandry, and the reform movement of the early 1870s supported ardently by urban and mercantile interests as well as farmers that sought, particularly in the Middle West, to bring railroad rates under public control. The two Movements overlap, but they were far from being one.

The National Grange of the Patrons of Husbandry was launched in 1867 by Oliver Hudson Kelley and a small group of government clerks. The organization was intended to be a secret society for farmers and their wives and was dedicated to serving their fraternal, social, educational, and economic needs. In part, Kelley may have intended to provide a national organization for farmer clubs, something others had urged in earlier decades. Kelley left the service of the federal government in 1868 to devote himself fully to organizing state and local granges, but the order grew slowly until 1872. In 1873 it was formally incorporated and passed into the hands of the members. Kelley remained as secretary, and Dudley W. Adams, master of the Iowa State Grange, was elected master of the National Grange. In 1874 a formal statement of purposes was adopted by the seventh annual session of the National Grange—a meeting that Solon J. Buck has characterized as the "most representative gathering of farmers which had ever taken place in the United States." At its

peak, in 1874, the Grange claimed nearly 22,000 local granges
with a total membership that has been estimated at more than a
million. In addition to their social, educational, and fraternal activities, state and local granges organized a large number of co-
operative ventures, some of which were successful, and many
members of the local granges, along with farmers enrolled in
other farmer organizations, supported the several reform, antimonoply, independent, and other political parties, prominent in
the early 1870s, that sought railroad and other reforms.

The reform movement that sought railroad rate and related
legislation was national in scope and drew support from many elements of society, but for many reasons it came to be viewed as
a revolt of the farmers against the railroads, and the warehouse
and elevator companies. Many farmers, probably most of them,
supported these reforms, although, as George H. Miller has
shown, the support was by no means restricted to farmers nor
did the farmers control the movement.

For a history of the National Grange, see Solon J. Buck, *The
Granger Movement. . .* Harvard, 1913) and for an excellent
account of the railroad-rate legislation in Iowa, Illinois, Wisconsin, and Minnesota, see George H. Miller, *Railroads and the
Granger Laws* Wisconsin, 1971) .

1 FROM *A Farmers' Platform Adopted at Centralia, Illinois, September 15, 1858*

This document represents the aspirations and grievances of many midwestern farmers in the 1850s and earlier. It also exhibits assumptions that were widely held about farmers and farming. The beliefs expressed about the need for unity of action and more knowledge, the primacy of farming in the Republic, the benefits of cooperatives, and the hostility toward "nonproducers" (middlemen), appeared again and again in various forms in the years after 1865.

We believe that the time has come when the producing classes should assert, not only their independence, but their supremacy; that non-producers can not be relied upon as guarantees of fairness; and that laws enacted and administered by lawyers are not a true standard of popular sentiment.

We believe that a general application to commerce of the principle that the majority should rule, would increase the income and diminish the outlay of producers, and, at the same time, elevate the standard of mercantile morality.

We believe that the producer of a commodity and the purchaser of it should, together, have more voice in fixing its price than he who simply carries it from one to the other.

We believe that the true method of guarding against commercial revulsions is to bring the producer and consumer as near together as possible, thus diminishing the alarming number and the more alarming power of non-producers.

We believe that in union there is strength, and that in union alone can the necessarily isolated condition of farmers be so strengthened as to enable them to cope, on equal terms with men

SOURCE. Jonathan Periam, *The Groundswell: A History of the Origins, Aims, and Progress of the Farmers' Movement* . . . (Cincinnati and Chicago, 1874) , pp. 204–206.

whose callings are, in their very nature, a permanent and self-
created combination of interests.

We believe that system of commerce to be the best which
transacts the most business, with the least tax on production, and
which, instead of being a master, is merely a servant.

We believe that good prices are as necessary to the prosperity
of farmers as good crops, and, in order to create such a power as
to insure as much uniformity in prices as in products, farmers
must keep out of debt; and that, in order to keep out of debt,
they must pay for what they buy and exact the same from others.

DECLARATION OF PRINCIPLES.

These truths we hold to be self-evident, that, as production
both precedes barter and employs more labor and capital, it is
more worthy the care and attention of governments and of individuals; that in the honorable transaction of a legitimate business
there is no necessity for secret cost-marks; that, in all well-regu-
lated communities, there should be the smallest possible number
of non-producers that is necessary to the welfare of the human
race; that labor and capital employed in agriculture should receive as much reward as labor and capital employed in any other
pursuit; that, as the exchanger is merely an agent between the
producer and consumer, he should not have a chief voice in the
establishment of prices; that the interests of agriculture and of
commerce can only be considered as identical when each has an
equal share in regulating barter; and that the principal road to
honor and distinction, in this country, should lead through productive industry."

PLAN OF OPERATIONS.

First. The formation of Farmers' Clubs wherever practical, the
object of which shall be to produce concert of action on all matters connected with their interests.

Second. The establishment, as far as possible, of the ready pay
system in all pecuniary transactions.

Third. The formation of wholesale purchasing and selling agencies in the great centers of commerce, so that producers may, in a great measure, have it in their power to save the profits of retailers.

Fourth. The organization of such a power as to insure the creation of a national agricultural bureau, the main object of which shall be an annual or semi-annual census of all our national products, and the collection and dissemination of valuable seeds, plants, and facts.

Fifth. The election of producers to all places of public trust and honor the general rule, and the election on non-producers the exception.

2 FROM *Jonathan Periam*
Illinois Farmers Demand State Control of Railroads, 1869

Although the call for the convention of farmers and the declaration adopted by the convention were intended to influence the constitutional convention about to convene in Springfield, both reflected views widely held by farmers in Illinois and other states in the Midwest. The constitutional convention met in Springfield in December 1869, and completed its work in May 1870. The new constitution, approved by popular vote in July, contained an article asserting the right of the legislature to regulate railroads. Businessmen, merchants, and many others as well as farmers supported this position (see Miller, Railroads, *Chapter 4).*

In March 1869, the Hon. Henry C. Wheeler, a farmer of Du Page County, Illinois, who had distinguished himself by his efforts to stir up his fellow-workers to a due sense of their power, issued a call for a convention of farmers of the North-west, to be

SOURCE. Jonathan Periam, *The Groundswell: A History of the Origins, Aims, and Progress of the Farmers' Movement* . . . (Cincinnati and Chicago, 1874) , pp. 224–230.

held at Bloomington, Illinois, on the following 20th of April. This document possesses great interest, as a forcible statement of the case of the producers against their enemies, and, still more, as leading to the first clearly-defined protest of an organization of farmers against the extortions of the monopolists, who were eating out their substance. Mr. Wheeler's manifesto was published in the principal papers of the North-west, and was as follows:

"THE TRANSPORTATION QUESTION"

"The Relation of the Carrying to the Producing Interest of the Country—Call for a Convention"

"*To the Farmers of the North-west.* Will you permit a working farmer, whose entire interest is identified with yours, to address to you a word of warning?

"A crisis in our affairs is approaching, and dangers threaten.

"You are aware that the price of many of our leading staples is so low that they cannot be transported to the markets of Europe, or even to our own seaboard, and leave a margin for profits, by reason of the excessive rates of transportation.

"During the war but little attention was given to the great increase in the price of freights, as the price of produce was proportionately high; but we look in vain for any abatement, now that we are obliged to accept less than half the former prices for much that we raise.

"We look in vain for any diminution in the carrying rates, to correspond with the rapidly-declining prices of the means of living, and of materials for constructing boats, cars, engines, and track; but, on the other hand, we see a total ignoring of that rule of reciprocity between the carrying and producing interests which prevails in every other department of trade and commerce.

Does it not behoove us, then, to inquire earnestly how long we can stand this descending scale on the one hand, and the ascending on the other, and which party must inevitably and speedily go to the wall?

"I by no means counsel *hostility* to the carrying interest—it is one of the producer's best friends; but, like the fire that cooks our food and warms our dwelling, it may also become the hardest of masters. The fire fiend laughs as he escapes from our control, and in an hour licks up and sweeps away the accumulations of years of toil.

"As we cherish the fire fiend, so we welcome the clangor of the carrier fiend as he approaches our dwellings, opening up communication with the busy marts of trade. But it needs no great stretch of imagination to hear also the cach! cach! cachinations of the carrier fiend as he speeds beyond our reach, and leaving no alternative but compliance with his exorbitant demands.

"Many of us are aware of the gigantic proportions the carrying interest is assuming. Less than forty years since the first railroad fire was kindled on this continent, but which now, like a mighty conflagration, is crackling and roaring over every prairie and through every mountain gorge. The first year produced fifteen miles; the last, 5,000.

"On the same mammoth scale goes on the work of organization and direction. By the use of almost unlimited means, it enlists in its service the finest talents of the land as officers, attorneys, agents, and lobbyists; gives free passes and splendid entertainments to the representatives of the people; and even transports whole legislatures into exceeding high mountains, showing them the kingdoms of the world, with lavish promises of reward for fealty and support: witness its land grants and franchise secured from the powers that be, such as no similar interest ever acquired even in the Old World.

"In Europe every corporation returns its franchises to the Crown within a specified time, while here their titles are more secure than the farmers' warranty deeds.

"Do you say that you are out of debt, and can stop producing when it does not pay?

"I tell you, my friends, that the carrying interest, with its present momentum unchecked, will soon acquire the power to tax your unincumbered possessions into leaseholds, and you and me into tenants at will.

"I fancy I hear the response: These things are so, but what can we do?

"Rather, my friends, what can we *not* do? What power can withstand the combined and concentrated force of the producing interest of this Republic? But what avails our strength if, like Polyphemus in the fable, we are unable to use it for want of eyesight; or, like a mighty army without discipline, every man fighting on his own hook; or, worse, reposing in fancied security while Delilahs of the enemy have well nigh shorn away the last lock of strength? In this respect we constitute a solitary exception, every other interest having long since protected itself by union and organization.

"As a measure calculated to bring all interested, as it were, within speaking distance, and as a stepping stone to an efficient organization, I propose that the farmers of the great North-west concentrate their efforts, power, and means, as the great transportation companies have done theirs, and accomplish something, instead of frittering away their efforts in doing nothing.

"And, to this end, I suggest a convention of those opposed to the present tendency to monopoly and extortionate charges by our transportation companies, to meet at Bloomington, Illinois, on the 20th day of April next, for the purpose of discussion, and the appointment of a committee to raise funds to be expended in the employment of the highest order of legal talent, to put in form of report and argument an exposition of the rights, wrongs, interests, and injuries with their remedies) of the producing masses of the North-west, and lay it before the authorities of each State and of the general government. Congress is now in session, and the Constitutional Convention of this State will then again be convened. Farmers, now is the time for action!

• • •

The various phases of the transportation and warehouse questions were earnestly discussed by the delegates, and as a rule, in an impartial spirit, only a small proportion of the speakers indulging in appeals to the passions of the audience. Strong resolutions were passed, denouncing the wrongs under which the producers labored, and the necessity of prompt and consecutive action to obviate these wrongs. A motion to send an official copy of the resolution to the President of the Constitutional Convention, then in session, was carried, and the following declaration was adopted:

"This Convention declares, First: That the present rates of taxation and transportation are unreasonable and oppressive, and ought to be reduced.

"Second: That our legal rights to transportation and market ought to be clearly set forth and defined.

"Third: That if there be any legal remedy under existing laws for the wrongs we suffer, such remedy ought to be ascertained and enforced.

"Fourth: That, if there be no such remedy, measures should be taken to secure one by appropriate legislation.

"Fifth: That statistics should be collected and published to show the relation of North-western products to those of the rest of the country.

"Sixth: That nothing can be accomplished for the enforcement of our rights, and the redress of our wrongs, without an efficient organization on the well-known principles that give the great corporations such tremendous power.

"Seventh: That, with honest-pay for honest labor, and compensation commensurate with great service, we can secure the assistance and support of the highest order of learning, ability, and skill.

"Eighth: That this Convention should appoint a commissioner of agricultural and carrying statistics, to prepare and publish, with the aid of eminent counsel, a report of the products of the North-west, the rights to market and transportation, and the remedies available for existing wrongs, the expenses thereof to be defrayed by subscription price for such report."

3 FROM   *Dudley W. Adams*
A Farmer Speaks About Farmers and Farming, 1872

Dudley W. Adams was born in Massachusetts in 1831. He had helped farm and had taught rural school before he moved to Iowa where he settled on government land near Waukon, in Allamakee county. He quickly became active in the county agricultural society and developed an impressive orchard on his farm. He joined the Iowa Horticultural Society and in 1868 became secretary of that body. In 1871, he helped organize the Iowa State Grange and was elected its master. The following year he was elected master of the National Grange, and in 1873 he was reelected to a three-year term under the new constitution. This speech, made before a joint meeting of the Granges of Union and Muscatine counties in October, 1872, seeks to tell farmers what they must do to improve their lot.

When railroad men have a convention, such persons as have had active experience in railroad business do the talking and have charge of the meeting.

Editorial conventions are attended by editors, and they, as firmly as any other class of people, are of the opinion that they are capable of managing their own business, and they are not in the habit of imploring the members of other callings to furnish the brains to amuse or instruct them.

Shoemakers have organized themselves into the order of St. Crispins, and consider themselves able to paddle their own canoe.

Lawyers not only feel competent to address and properly edify conventions of their own profession, but their modesty does not forbid them from rendering valuable assistance to less favored classes by a free use of their surplus talent.

SOURCE: Edward Winslow Martin (pseudonym, James Dabney McCabe) *History of the Grange Movement; or, The Farmer's War Against Monopolies . . . with A History of the Rise and Progress of the Order of Patrons of Husbandry . . .* (Philadelphia, Chicago, Cincinnati, pp. 515 ff. Reprinted 1968, Burt Franklin, New York).

But, when the tillers of the soil have met in an agricultural society of any kind, it has been usually customary to select a lawyer, doctor, editor, or politician to tell us what he knows about farming. The idea has very rarely occurred to the managers of such institutions that it might be possible for a farmer to have anything to say on such occasions which should be either appropriate, interesting, or instructive. When these professional oracles of our professional managers' selection open their mouths, we are edified with a rehash of such ideas as may be prevalent in the community, served up in a great variety of forms, and presented in a great many different and beautiful lights, depending for its coloring upon the business of the orator, as this is the standpoint from which we are viewed, and, of course, this view determines the nature of the picture. Lawyers and doctors in beautiful colors paint the nobleness and independence of the farmer's life. They tell us we are the most intelligent, moral, healthy, and industrious class in all the land, and all our present is calm and our future happy. Merchants tell us that no business is so sure and free from care as farming, and that in no other calling do so few men end in bankruptcy. Politicians laud in stentorian tones the "honest yeomanry," "the sinews of the land," the "bulwarks of our nation's liberties," "the coarse blouse of homespun which covers the true and honest heart," and deluges more of equally fulsome and nauseating stuff.

Soft-handed agricultural editors give long dissertations on the necessity of saving all the spare moments, and converting them into some useful purpose. They tell us how rainy days may be laboriously used in mending old rake-handles, and winter evenings utilized by pounding oak logs into basket stuff, while our wives and daughters can nobly assist in averting bankruptcy by weaving the baskets or ingeniously making one new lamp-wick out of the remains of three old ones.

It has never occured to these very wise instructors that farmers and farmers' families are human beings, with human feelings, human hopes and ambitions, and human desires. It will doubtless be a matter of surprise for them to learn that farmers may possibly entertain some wish to enjoy life, and have some other object in life besides everlasting hard work and accumulating a few paltry dollars by coining them from their own life-blood, and stamping them with the sighs of weary children and worn wives.

I tell you, my brother tillers of the soil, there is something in this world worth living for besides hard work. We have heard enough of this professional blarney. Toil is not in itself necessarily glorious. To toil like a slave, raise fat steers, cultivate broad acres, pile up treasures of bonds and lands and herds, and at the same time bow and starve the godlike form, harden the hands, dwarf the immortal mind, and alienate the children from the homestead, is a damning disgrace to any man, and should stamp him as worse than a brute.

I will be met right here with the thousand time repeated rejoinder, "Oh, we farmers have to work hard. We can't get along as mechanics in town do with ten hours' work. We can't afford to hire help. We can't afford to have holidays. We can't get time to make a vegetable, flower, and fruit garden, and supply our wants with vegetables, flowers, and fruits. We can't get time to make a lawn and plant trees around the house." You can't? You can't? Then what are you farming for? As men, as citizens, as fathers, as husbands, you have no right to engage in a business which will comdemn yourself and your dependents to a life of unrewarded toil. If the calling of agriculture will not enable you and yours to escape physical degradation, and mental and social starvation; if it does not enable you to enjoy the amenities, pleasures, comforts, and necessities of life as well as other branches of business, it is your duty to abandon it at once, and not drag down in misery your dependent family. But I do not believe we need be driven to this alternative. I *do* believe that agriculture, followed as a business, with a reasonable regard to business principles, can be made a business success. I believe that by keeping steadily in view the primary end of life—our happiness, our comfort, our bodily health, our mental improvement and growth —they can be as well attained or better than in any other calling. Right here is the great difficulty; right here with ourselves is the remedy: We *work* too much and *think* too little. We make our hands too hard, while our brains are too soft. The day is long past when muscle ruled the world. Brain is the great motive power of this age, and muscle but a feeble instrument. The locomotive, tearing along, jarring the earth below, outstripping the wind above, and bearing in its train the beauty, honor, and treasure of a State, represents brains. The dusty, sweaty footman, wearily plodding along, carrying a pack on his back, symbolizes muscle. The self-raking reaper, driven with gloved and unsoiled hands, sweeping down like a fable the golden grain, represents brains. The bowed husbandman, painfully gathering handfuls of straw and cutting them with a sickle, represents muscle. The steamboat, plowing its way with ease against the strongest current of our swift and noble rivers, is brains. The dug-out, slowly creeping along the willow-margined shore, propelled by the Indian's paddle, is muscle. The sewing-machine, which stitches faster than the eye can follow and never eats or tires, is brains. The weary, pale, and worn wife, painfully toiling over the midnight task, is muscle. How futile the attempt, then, for muscle to compete against mind in the great battle of life! A wise man once wrote, "The wisdom of a learned man cometh with opportunity of leisure"; and in that sentence is food for reflection and thought sufficient for an entire sermon. Unless farmers devote more time to the use of the brain and the improvement of the mind, and less to wearying and exhausting muscular labor, how can they hope to successfully compete against the vigorous minds of the present age? It is not the skilful hand, the strong arm, or the watchful eye alone that will in these days bring success to the farmer. These are needful, but a cultivated, intelligent, active brain to direct them is of ten times more importance.

Let us take a candid look at the situation, and see if we cannot discover what is the matter. Let us try and see if there is any good reason why the great majority should be governed and oppressed by a small minority.

In human affairs effects follow causes; results are accomplished by action, even when the actors are unseen. Look at our State and national Governments, and who are the men to whom we entrust this great responsibility? Look at our boards of trade, industrial expositions, and in fact any great project for the advancement of science, art, liberty, or industry, and you will find at its head and the moving spirit thereof a lawyer, doctor, preacher, student, merchant, or, in fact, almost anything but a farmer. These men rule the nation. They shape the laws; they make the channels of trade, and place trade in its channels. They build ships, harness steam to their wagons, make lightning carry their messages; they compel rivers to turn their saws, twirl their spindles, and throw their shuttles. They use their brains, and mind governs the world.

Just think of competing against such men by stupidly hoeing corn fifteen hours a day and selling it at twenty cents a bushel, and then laying awake nights, growling at railroad men and merchants. The dog who barks at the moon comes nearer accomplishing his purpose than such a growler. Why have not farmers taken a position of influence and power in the councils of the nation and otherwise, in proportion to their numbers and wealth? Simply because we have not used our brains.

I speak it in sorrow; I admit it with deep and burning shame, that the farmers can furnish but comparatively few men whose minds are fitted to organize great enterprises. Look at the farmers in our Legislature. In numbers they are very small in proportion to the population of the State, and smaller yet in the influence they have upon the legislation. When they come in contact with men who are in the habit of close and logical reasoning, they, with a few exceptions, prove wanting. It may, and probably will be said that headwork will not hoe corn or feed the pigs. Granted. But prove to me that an intelligent man is disqualified from performing the duties of a farmer and you will prove to me that farming is a business which it is disgraceful to follow, and that it is grossly unjust to say aught to induce any young man of common sense to become a farmer.

. . . If agriculture will give scope to thought and research; if it will cause a man to think while he works and study while he has leisure; if his business is such that talent and tact will transform his soil to gold and his house into a beautiful and happy home; if the same amount of bodily and mental labor on the farm will produce as much pleasure, wealth, and happiness as in the shops, counting-room, and mines, then we may conscientiously recommend agriculture as one of the desirable employments. Can this be done?

Brother Patrons of Husbandry, our Order has been formed to assist in answering this great question in the affirmative. How shall we proceed?

I do not underrate the importance of making an effort to buy our reapers a few dollars cheaper and sell our wheat a few cents higher and get our freights a little lower. What is gained in this way is certainly added to the profits of the farm, but I very much fear that many members of the Order place too high a value upon this matter of purchase and sale. This is not what ails us. It does not reach the root of the difficulty at all. It only prunes away a few slender twigs which grow again in a single night. We can never accomplish what we want, and make agriculture respectable, remunerative, and desirable; farmers intelligent, contented, and honored; farmers' wives envied and respected, and farmers' sons and daughters eagerly sought by the wise, good, learned, and beautiful of the land for husbands and wives; we cannot make beautiful homes, fertile farms, and improving flocks by saving five dollars on a plow and five cents a bushel on wheat. No! Never! When we build like that we must dig deeper, lay the foundations broader, and use *brains* as the chief stone of the corner. An ox excels us in strength, a horse in speed. The eagle has keener sight, the hare a quicker ear, the deer a finer sense of smell; but man excels them all in mind and rules above them all. So among men it is not the strong, the swift, the keen-sighted, the quick-eared or fine-scented who rules the world, but the clear-headed. Human beings are like pebbles on the sea shore; by rubbing against each other they become rounded, smooth, polished, symmetrical: alone, they are rough, uncouth, repulsive.

Farmers are too much alone. We need to meet together to rub off the rough corners and polish down into symmetry. We want to exchange views, and above all we want to learn to think. A man who has performed fourteen hours of severe physical labor is in no condition to think, and we may as well decide at once that any class of men which starts out in life by working at severe labor fourteen hours of the twenty-four, and faithfully adheres to the practice, will fill forever the position of hewers of wood and drawers of water for men who use the God-given mind, and nourish the soul with liberal and abundant mental food.

4   FROM   *The Constitution of the National
Grange of the Patrons of Husbandry, 1873*

*Although launched in 1867 by Oliver Hudson Kelley and his
associates, the Grange was not incorporated until 1873. Until
that time, and during the early period of rapid growth in 1872, it
continued to be managed mainly by O. H. Kelley. The constitution of 1873 rewards careful reading, since it reflects with fair
accuracy what the leaders thought the organization should be
and do. It would retain its ritual; it would be open to all persons
of either sex interested in agricultural pursuits; it would collect
agricultural and other statistics; it would distribute useful information to its members; it would not neglect the sick among its
members; it would oppose "wanton cruelty" to animals; and it
would explicitly forbid discussion of political and religious questions, and also would forbid religious and political tests for membership.*

PREAMBLE TO THE CONSTITUTION

Human happiness is the acme of earthly ambition. Individual
happiness depends on general prosperity.

The prosperity of a nation is in proportion to the value of its
productions.

The soil is the source from whence we derive all that constitutes wealth; without it, we would have no agriculture, no manufactures, no commerce. Of all the material gifts of the Creator,
the various productions of the vegetable world are of the first importance. The art of agriculture is the parent and precursor of all
arts, and its products the foundation of all wealth.

The productions of the earth are subject to the influence of
natural laws, invariable and indisputable; the amount produced

SOURCE. Jonathan Periam, *The Groundswell: A History of the Origin,
Aims, and Progress of the Farmers' Movement* . . . (Cincinnati and Chicago,
1874) , pp. 168–172.

will, consequently, be in proportion to the intelligence of the producer, and success will depend upon his knowledge of the action of these laws, and the proper application of their principles.

Hence, knowledge is the foundation of happiness.

The ultimate object of this organization is mutual instruction and protection, to lighten labor by diffusing a knowledge of its aims and purposes, to expand the mind by tracing the beautiful laws the Great Creator has established in the universe, and to enlarge our views of Creative wisdom and power.

To those who read aright, history proves that in all ages society is fragmentary, and successful results of general welfare can be secured only by general effort. Unity of action can not be acquired without discipline, and discipline can not be enforced without significant organization; hence, we have a ceremony of initiation, which binds us in mutual fraternity as with a band of iron; but although its influence is so powerful, its application is as gentle as that of the silken thread that binds a wreath of flowers.

The Patrons of Husbandry consist of the following:

CONSTITUTION

ARTICLE I.—*Officers.*—Section 1. The Officers of a Grange, either National, State, or Subordinate, consist of and rank as follows: Master, Overseer, Lecturer, Steward, Assistant Steward, Chaplain, Treasurer, Secretary, Gate-Keeper, Ceres, Pomona, Flora, and Lady Assistant Steward. It is their duty to see that the laws of the Order are carried out.

Sec. 5. The officers of the respective Granges shall be addressed as "Worthy."

ARTICLE IV.—*Ritual.*—The Ritual adopted by the National Grange shall be used in all Subordinate Granges, and any desired alteration in the same must be submitted to, and receive the sanction of, the National Grange.

ARTICLE V.—*Membership.*—Any person interested in agricultural pursuits, of the age of sixteen years (female), and eighteen years (male), duly proposed, elected, and complying with the rules and regulations of the Order, is entitled to membership and the benefit of the degrees taken. Every application must be accompanied by the fee of membership. If rejected, the money will be refunded. Applications must be certified by members, and balloted for at a subsequent meeting. It shall require three negative votes to reject an applicant.

ARTICLE VIII.—*Requirements.*—Section 1. Reports from Subordinate Granges relative to crops, implements, stock, or any other matters called for by the National Grange, must be certified to by the Master and Secretary, and under seal of the Grange giving the same.

Sec. 2. All printed matter, on whatever subject, and all information issued by the National or State to Subordinate Granges, shall be made known to the members without unnecessary delay.

Sec. 3. If any brothers or sisters of the Order are sick, it shall be the duty of the Patrons to visit them, and see that they are well provided with all things needful.

Sec. 4. Any member found guilty of wanton cruelty to animals shall be expelled from the Order.

Sec. 5. The officers of Subordinate Granges shall be on the alert in devising means by which the interests of the whole Order may be advanced; but no plan of work shall be adopted by State or Subordinate Granges without first submitting it to, and receiving the sanction of, the National Grange.

ARTICLE XII.—*Restrictions.*—Religious or political questions will not be tolerated as subjects of discussion in the work of the Order, and no political or religious tests for membership shall be applied.

5   FROM *The Illinois Farmers' Association, 1873*

The constitution of the Illinois State Farmers' Association and the resolutions offered to the convention that brought the association into being, exhibit the broad spectrum of farmer concerns in the panic year of 1873. The Illinois Farmers' Association represents one of the fairly successful attempts in a midwestern state to bring all farmer organizations together to advance farmers' interests. It is worth noting that in Illinois the farmers' clubs were more numerous than the local granges.

1803995
THE CALL FOR THE CONVENTION

In compliance with the duty assigned them, the Executive Committee appointed at Kewanee issued a call for a State Convention, to be held at Bloomington, on the 15th and 16th days of January 1873, of which call the following is the more essential portion:

Farmers' Convention

"*Equal and Exact Justice to All; Special Privileges to None.*" The undersigned, the Executive Committee appointed by the Convention of delegates from Farmers' Clubs, held at Kewanee, Oct. 16th and 17th, 1872, in pursuance of the duties assigned them, do hereby invite each Farmers' Club, Grange, or other agricultural, horticultural, or industrial association of the State of Illinois, to send delegates for every thirty-three members, and fraction in excess of that number, (*Provided,* That every organization shall be entitled to at least one delegate), to an Illinois Farmers' State Convention, to be held in the City of Bloomigton, Wednesday and Thursday, Jan. 15 and 16, 1873, commencing at

SOURCE. Jonathan Periam, *Groundswell: A History of the Origins, Aims, and Progress of the Farmers' Movement* . . . (Cincinnati and Chicago, 1874) pp. 243–244, and 256–262.

9 A.M. on Wednesday, with three sessions each day—at 9 A.M., 2
P.M., and 7 P.M.

The purposes of said Convention is to perfect the organization
made at Kewanee, by the formation of a State Farmers' Association from said delegates, adoption of a constitution, and for securing the organization and representation of associations in every county, and, if possible, in every township, of the State; to
discuss and insist upon reform in railway transportation, the sale
of agricultural implements, the sale of farm products by commission merchants, and such other abuses as have grown up in our
midst, and are now taxing and impoverishing producers and consumers; and to transact such other business as may be brought
before the Convention.

ARTICLE I. This organization shall be known as the Illinois
State Farmers' Association.

ARTICLE II. Its object shall be the promotion of the moral, intellectual, social, and pecuniary welfare of the farmers of Illinois.

ARTICLE III. Its members shall consist of delegates from the
various Farmers' Clubs, Granges, and Agricultural and Horticultural Societies of the State, each of which shall be entitled to at
least one delegate, and where the number of its members exceed
fifty, to one delegate for every one hundred members or fraction
exceeding half that number. The members of the State Board of
Agriculture shall be, *ex-officio,* members of this association, and
from counties or parts of counties in which Clubs, Granges, or
other Agricultural or Horticultural Societies are not organized,
persons, not delegates, may be admitted by vote of this association. All members shall pay an annual fee of $1.

THE RESOLUTIONS

The Committee on Resolutions reported as follows:

WHEREAS, The Constitution of Illinois requires the Legislature
to pass laws to correct abuses and prevent unjust discrimination
and extortion by railroads; and,

WHEREAS, The Legislature has complied with this provision of
the Constitution; and,

WHEREAS, The Railroads in the State of Illinois stand in open defiance of the laws, by charging rates greatly in excess of what the laws allow, and by unjust discriminations and extortions; and,

WHEREAS, These exactions and extortions bear most heavily upon the producing classes; therefore, be it

Resolved, That this convention of farmers and producers insist upon the enforcement of these laws.

Resolved, That in obedience to the universal law that the creature is not above the creator, we declare our unalterable conviction that all corporations are subject to regulation by law.

Resolved, That we call upon every department of the State government—the executive, legislative, and judicial—in their joint and several capacities, to execute the Constitution and laws now in force; and if amendments or new laws are needed to enforce obedience, we call for their speedy enactment.

Resolved, That cheap transportation is of vital interest to the West, and that every combination to increase the price above what is just and legitimate is a conspiracy against the rights of the people, and a robbery which we loudly protest against.

Resolved, That in the efforts of our officers to execute the laws in question, no narrow policy should be pursued by the Legislature, but that the magnitude of the matter at stake demands that ample appropriations be made, to enable those in charge of the object to act with vigor and effect.

Resolved, That the power of this, and all local organizations, should be wielded at the ballot-box by the election to all offices, from highest to lowest—legislative, executive, and judicial—of such, and only such, persons as sympathize with us in this movement, and believe, as we do, that there is a rightful remedy for this wrong, and that it can and must be enforced; and to this end we pledge our votes at all elections where they will have a bearing against the wrong in question.

Resolved, That the late decision in the McLean County Circuit Court, sustaining the constitutionality of our railroad law, is sound, and we hail it with satisfaction.

Resolved, That persons traveling upon the railroads of the State, having tendered to the conductor the legal fare, are in line of their duty, and as they have complied with all their legal obligations, are entitled to the protection of the civil power of the

State; and any conductor or other officer or employee of the road attempting to disturb any such person, or eject him from the cars, are violators of the peace and dignity of the State, and should be punished by exemplary penalties.

The railroad question was again brought up, by Mr. Stephen R. Moore, of Kankakee, who reviewed the subject, and offered the following resolutions, which were referred to the appropriate committee. They are as follows:

Resolved, That all transportation companies, lines, and persons shall have the right to run their cars, said roads paying as toll therefor such compensation as the Board of Directors shall determine upon.

Resolved, That the Board of Directors for the management of the said roads shall be elected by the lower House of Representatives of the State named as corporators.

Resolved, That one Senator and Representative be requested to appoint three persons who shall be empowered to proceed to ·the Legislatures of States through which the road will pass, and ask co-operation by the States, and request that each State will appoint three persons, who shall constitute a joint committee.

Resolved, That the charter for the constitution of such national railway should be granted by the national government, and said railway shall forever remain a public highway.

Resolved, That the States through which this national freight road shall be built shall become corporators under said charter.

Resolved, That in the States through which, and along which, the road should be built, the people thereof shall construct the same through the territory of said States respectively.

Resolved, That said railway shall ever remain under the control of said States, which States shall exercise the power of regulating the tariff rates.

Resolved, That it is the sense of this convention that a double track, steel rail, freight railway should be built from Lincoln, Neb., in the West, crossing the States of Iowa and Missouri as near upon a line as may be, running thence East on air line to Youngstown, near the eastern boundary of Ohio; thence following the Gardner survey to some point in Pennsylvania, to be determined upon hereafter; from thence with three diverging lines running to New York, Philadelphia, and Baltimore; and that said railway should be used exclusively for a freight line. The committee to prepare a charter for the construction of the road, to submit it to the respective State Legislatures for approval, and, when approved, to present the same to Congress, and ask for its passage.

Resolved, That the chairman appoint a committee of five to present these resolutions to the Legislature, and ask that body to carry out their spirit.

Mr. Carter, from the Committee on Resolutions, submitted a report, as follows:

Resolved, That we recommend to our Legislature the enactment of a law making it a misdemeanor for any county or State officer to accept a free pass from any railroad, while holding office.

Resolved, That we view with favor the opening of feasible water communications, and all propositions to so improve and enlarge the great water line of the lakes and the St. Lawrence as to practically bring tide water to Chicago; and, for this purpose, completing the Illinois river improvement, and the extension of the canal to Rock Island, so as to connect the vast interior river system with the ocean commerce at our great commercial city, meet our approbation.

Resolved, That it is the sense of this convention that in the appointment of Railroad and Warehouse Commissioners, at least one of the members of that commission should be a man whose business interests, sympathies, and knowledge of the experiences and wants of the farmer class, should fairly constitute him a representative man of that class, and who shall be so recommended by them.

Resolved, That in order to accomplish the ends arrived at by this Convention, we earnestly recommend the organization of the farmers throughout the State into Clubs, and Granges of Patrons of Husbandry.

The following resolution was offered and adopted:

Resolved, That this Convention appoint Capt. J. H. Rowell and R. W. Benjamin to proceed to our Legislature, and procure an act condemning all railroads that are running in violation of the law, and we further recommend that a commission be appointed to take charge of such road or roads, and run them in compliance with the law.

The following miscellaneous resolutions were offered and referred to the Committee on Resolutions:

Resolved, That the Legislature of this State be requested to instruct our Senators, and request our Representatives in Congress, in view of the depressed condition of the agricultural interests of this State and all others dependent thereon, except that of railway transactions, to insist upon the utmost economy in appropriations and frugality in expenditure of national moneys consonant with the necessities of the country.

Resolved, That we are in favor of removing the duties on iron, lumber, and salt.

Resolved, That farmers buy no implements of those manufacturers or their agents who have entered into any conspiracy agreeing not to sell their implements to Farmers' Associations.

• • •

6   FROM   *Edward W. Martin*
The Declaration of Purposes of the Patrons of
Husbandry, 1874

At the first annual session of the Grange after it had "gone public"—the seventh in the official chronology—a declaration of purposes was adopted and published. Little in this document supports the position sometimes advanced by scholars—that the Patrons of Husbandry were atavistic in their economic notions and fiercely opposed to industrialization. Although Buck asserts that the declaration was written by J. W. A. Wright of California with the help of a committee, the sentiments of the declaration

SOURCE. Edward Winslow Martin (pseudonym, James Dabney McCabe) *History of the Grange Movement; or, the Farmer's War Against Monopolies . . . with A History of the Rise and Progress of the Order of Patrons of Husbandry . . .* (Philadelphia, Chicago, Cincinnati, pp. 535–539. Reprinted 1968, Burt Franklin, New York) .

are in accord with the views expressed in 1872 by Dudley W. Adams, Master of the National Grange, 1873 to 1876. The annual session of 1874 also proposed several changes in the constitution. One of the changes provided that membership was open to "any person engaged in agricultural pursuits and having no interest in conflict with our purposes. . ." (see Buck, The Granger Movement. . ., pp. 62 ff).

Profoundly impressed with the truth that the National Grange of the United States should definitely proclaim to the world its general objects, we hereby unanimously make this Declaration of Purposes of the Patrons of Husbandry:

1. United by the strong and faithful tie of Agriculture, we mutually resolve to labor for the good of our order, our country, and mankind.

2. We heartily indorse the motto, "In essentials, unity; in non-essentials, liberty; in all things, charity." We shall endeavor to advance our cause by laboring to accomplish the following objects;

To develop a better and higher manhood and womanhood among ourselves. To enhance the comforts and attractions of our homes, and strengthen our attachments to our pursuits. To foster mutual understanding and cooperation. To maintain inviolate our laws, and to emulate each other in labor. To hasten the good time coming. To reduce our expenses, both individual and corporate. To buy less, and produce more, in order to make our farms self-sustaining. To diversify our crops, and crop no more than we can cultivate. To condense the weight of our exports, selling less in the bushel, and more on hoof and in fleece. To systematize our work and calculate intelligently on probabilities. To discountenance the credit system, the mortgage system, the fashion system, and every other system tending to prodigality and bankruptcy.

We propose meeting together, talking together, working together, buying together, selling together, and in general acting together for our mutual protection and advancement as occasion may require. We shall avoid litigation as much as possible by arbitration in the Grange. We shall constantly strive to secure entire harmony, good will, vital brotherhood among ourselves, and to make our order perpetual. We shall earnestly endeavor to suppress personal, local, sectional, and national prejudices, all unhealthy rivalry, all selfish ambition. Faithful adherence to these principles will insure our mental, moral, social, and material advancement.

3. For our business interests we desire to bring producers and consumers, farmers and manufacturers, into the most direct and friendly relations possible. Hence we must dispense with a surplus of middlemen; not that we are unfriendly to them, but we do not need them. Their surplus and their exactions diminish our profits. We wage no aggressive warfare against any other interests whatever. On the contrary, all our acts and all our efforts, so far as business is concerned, are not only for the benefit of the producer and consumer, but also for all other interests, and tend to bring these two parties into speedy and economical contact. Hence we hold that transportation companies of every kind are necessary to our success; that their interests are intimately connected with our interests, and harmonious action is mutually advantageous. Keeping in view the first sentence in our declaration of principles of action, that "individual happiness depends upon general prosperity," we shall therefore advocate for every State the increase in every practicable way of all facilities for transporting cheaply to the seaboard, or between home producers and consumers, all the productions of our country. We adopt it as our fixed purpose to open out the channels in Nature's great arteries, that the lifeblood of commerce may flow freely. We are not enemies of railroads, navigation, and irrigating canals, nor of any corporation that will advance our industrial interests, nor of any laboring classes. In our noble order there is no communism, no agrarianism. We are opposed to such spirit and management of any corporation or enterprise as tends to oppress the people and rob them of their just profits. We are not enemies of capital, but we oppose the tyranny of monopolies. We long to see the antagonism between capital and labor removed by common consent and by enlightened statesmanship worthy of the nineteenth century. We are opposed to excessive salaries, high rates of interest, and exorbitant profits in trade. They greatly increase our

burdens, and do not bear a proper proportion to the profits of producers. We desire only self-protection and the protection of every interest of our land by legitimate transactions, legitimate trade, and legitimate profits.

4. We shall advance the cause of education among ourselves and for our children by all just means within our power. We especially advocate for our agricultural and industrial colleges that practical agriculture, domestic science, and all the arts which adorn the home be taught in their courses of study.

5. We especially and sincerely assert the oft-repeated truth taught in our organic law, that the Grange, National, State, or subordinate, is not a political or party organization. No Grange, if true to its obligations, can discuss political or religious questions, nor call political conventions, nor nominate candidates, nor even discuss their merits in its meetings. Yet the principles we teach underlie all true politics, all true statesmanship, and if properly carried out will tend to purify the whole political atmosphere of our country. For we seek the greatest good to the greatest number, but we must always bear in mind that no one by becoming a Patron of Husbandry gives up that inalienable right and duty which belongs to every American citizen, to take a proper interest in the politics of his country. On the contrary, it is right for every member to do all in his power legitimately to influence for good the action of any political party to which he belongs. It is his duty to do all he can in his own party to put down bribery, corruption and trickery; to see that none but competent, faithful and honest men, who will unflinchingly stand by our industrial interests, are nominated for all positions. It should always characterize every Patron of Husbandry that the offices should seek the man and not the man the office. We acknowledge the broad principle that difference of opinion is no crime, and hold that progress towards truth is made by differences of opinion, while the fault lies in bitterness of controversy. We desire a proper equality, equity, and fairness, protection for the weak, restraint upon the strong; in short, justly distributed burdens and justly distributed power. These are American ideas, the very essence of American independence, and to advocate the contrary is unworthy of the sons and daughters of an American republic. We cherish the belief that sectionalism is, and of right should be, dead

and buried with the past. Our work is for the present and the future. In our agricultural brotherhood and its purposes, we shall recognize no North, no South, no East, no West. It is reserved by every Patron, as the right of a freeman, to affiliate with any party that will best carry out his principles.

6. Ours being peculiarly a farmers' institution, we cannot admit all to our ranks. Many are excluded by the nature of our organization, not because they are professional men, or artisans, or laborers, but because they have not a sufficient direct interest in tilling or pasturing the soil, or may have some interest in conflict with our purposes. But we appeal to all good citizens for their cordial cooperation to assist in our efforts toward reform, that we may eventually remove from our midst the last vestige of tyranny and corruption. We hail the general desire for fraternal harmony, equitable compromise, and earnest cooperation, as an omen of our future success.

7. It shall be an abiding principle with us to relieve any of our suffering brotherhood by any means at our command. Last, but not least, we proclaim it among our purposes to inculcate a proper appreciation of the abilities and sphere of woman, as is indicated by admitting her to membership and position in our Order. Imploring the continued assistance of our Divine Master to guide us in our work, we here pledge ourselves to faithful and harmonious labor for all future time to return by our united efforts to the wisdom, justice, fraternity, and political purity of our forefathers.

7 FROM   *Willard G. Flagg*
An Illinois Farmer Talks About the Farmers'
Movement, 1874

Willard G. Flagg, described by Miller as "one of the wealthiest commercial farmers in the state" (of Illinois), was president of one of the most active of the farmer organizations of the period. This paper was prepared at the invitation of the Social Science Association of Boston. Flagg here offers a farmer's view of the farmers' problems in the mid-1870s.

THE FARMERS' MOVEMENT IN THE WESTERN STATE

A circular sent out over a year since by the Department of Agriculture elicited from returns from over half the counties of the Union, the fact that a small minority of the reports expressed no discouragement. "The leading difficulty in the West," says the statistician, "and a prominent one in all sections, is the burden of railroad transportation." Middle men, and the monopolies formed by their combination, are pronounced a "terrible scourge," especially in some of the Atlantic States. Upon these two abused functions of trade and transportation, the agricultural population instinctively fixed, and subsequent discussion has shown they were not far wrong.

From these as immediate causes, I believe, the Farmers' Movement began. It appropriated such existing organizations of the agricultural class as seemed best adapted to its purpose. The agricultural societies, or boards of agriculture, whose chief function was the holding of an annual exhibition, were generally passed by; but the Farmers' Clubs, organized for discussion and mutual instruction in the art of agriculture, and the Granges of Patrons of Husbandry, whose original object was educational and social, came nearer what was needed, and were appropriat-

SOURCE. *Journal of Social Science,* 6:100–113 (July 1874) .

ed, multiplied, and turned upon the new work. The loose organization and early mortality of the clubs, and the secrecy and pro-
hibition of political discussion in the granges, were drawbacks to
their efficiency; but in spite of them organization has gone on
rapidly. In nearly every State of the Union enough local granges
have been organized to make the organization of a State Grange
admissible. Probably one million persons belong to that order
alone. On the first day of January, 1874, they were strongest in
the order named in Iowa, Missouri, Indiana, Illinois, Kansas,
Mississippi, Georgia, and Minnesota, none of which had less
than four hundred granges. The number of clubs is less attaina-
ble. In Illinois, on the first of January, there were not far from
nine hundred with a membership of forty-five thousand persons.

The example of these rapidly forming organizations reacts
upon the labor organizations; and we find them increasing their
numbers and changing their form to something analagous to the
Patrons of Husbandry. The old trades-unions are strengthened,
or, what is better, tend to become more fused into a class organization, with broader and less selfish purposes. Sooner or later all
these industrial classes promise to join hands in the common
name of *Labor,* whose rights, heretofore antithesised against capital, must rather be stated as Lincoln put the case in 1861: "Labor is prior to and independent of capital. Capital is only the
fruit of labor, and never could have existed if labor had not first
existed. Labor is the superior of capital, and deserves much the
higher consideration." If this be true, it is evident that the legislatures, the executive officers, and the judiciary of our States and
nation, have heretofore spent an undue amount of time in incor-
porating, looking after, and doing the business of capital, in the
shape of railroad, banking, insurance, and other corporations devised for the purpose of enabling privileged classes to average
higher profits than common men could individually gain. And it
is evident that a government based on the idea of the political
equality of all men is by no means performing its proper function
in building up a plutocracy. Thus the movement which began
with the agricultural class gives new life and energy to our labor
reforms, and promises to become a general, movement in the
direction of political reform, independent of all existing political
organizations.

The discussions which ensue in ferreting out the causes of existing ills are most valuable in exposing abuses and educating the people to more radical theories than have yet been accepted by those who have thus far controlled legislation. In the light of these discussions further insight is obtained into the more remote causes of the Farmers' Movement. Many of these equally affect all the industrial classes, and some of them call for the attention of all fair-minded men of whatever pursuit in life. They are by no means class grievances—although even a class grievance affecting from half to three quarters of the population is a serious affair,—but national disasters, in the highest degree detrimental to the producing classes on whose prosperity the welfare of the country depends. I will enumerate some of them.

I. The wealth of the country, although the product of the labor of our industrial classes, in great measure, does not remain in the hands of those classes, but accumulates in the hands of a relatively small number of non-producers. Compare the increase of the wealth engaged in different investments from 1850 to 1870 in the United States, according to the census:

	1850	1860	1870
Total wealth	$7,135,780,228	$14,159,616,068	$30,068,518,507
Agricultural wealth	3,967,343,580	7,980,493,063	11,124,958,737
Manufacturing wealth	533,245,351	1,009,855,705	2,118,208,769
Mining wealth	—	—	222,384,854
Fisheries	—	—	7,469,575
Total industrial wealth	4,500,588,931	8,990,348,778	13,473,021,935
Total other wealth	2,635,191,297	5,169,267,290	16,585,496,572

Assuming these figures to be sufficiently correct for our purpose, we notice immediately a wonderful difference in the rate of increase of industrial as compared with other wealth during the last decade. In the decade 1850–60 the various classes of wealth enumerated increased at very nearly the same rate. But in the period between 1860 and 1870 the agricultural wealth of the country increased but 40 per cent, or 4 per cent per annum, and the manufacturing and mining wealth 110 per cent., or 11

per cent. per annum. The increase of all the industrial wealth
was about 50 per cent., or 5 per cent. per annum, while the
wealth of the country not industrial increased 200 per cent., or
20 per cent. per annum, or 4 and 5 to 1 as compared with the
industrial and agricultural wealth. This statement may be quali-
fied by the fact that the industrial classes have considerable
amounts invested outside of their occupation; that there is more
"inflation" in the estimated wealth not agricultural, etc. Yet the
undeniable fact remains that the industries of the country receive
no fair share of their earnings.

The actual *production* of wealth by the industrial classes, and
its percentage on the capital invested is far greater, than the
above figures would indicate.

	Capital Invested	Persons Employed	Earnings 1870	Earnings per Head	Earnings per cent of capital
Agriculture	$11,124,958,73	5,922,471	$2,447,538,628	$413	22
Manufactures	2,118,208,769	2,053,966	1,743,898,200	849	62
Mining	222,384,854	154,328	138,323,303	896	82
Fisheries	7,469,575	20,504	1,642,276	461	22
	$13,473,021,935	8,151,296	$4,331,402,437	$531	32

Two facts are here significant—the relatively and absolutely
low earnings of agricultural labor, and the fact that low as they
are they are far above the actual increase of wealth in agriculture. If a comparison could be made of industrial labor with professional and other occupation the ratio would of course be
much smaller.

All this constitutes a most serious charge against the political
economy of a country in which such things are possible. Think of
the facts and the necessary consequences if such a state of affairs
be permitted to go on! Agricultural wealth in a new and fertile
country, increasing at the rate of 4 per cent. per annum, while
the non-producing wealth increases at the rate of 20 per cent.!
How long can such a course be run and not renew the ancient
story of serfdom, or familiarizing America with the ignorance,
degradation, and suffering that disgrace the English agricultural
districts to-day?

II. The conditions and often the direct causes of this extraordinary difference in prosperity as compared with the nearly uniform rate of increase between 1850 and 1860 are, I believe, mainly legislative. The more cunning, unscrupulous, and corrupting hand of corporation and capitalist has moulded legislation to its will in its own interest. It has been quite possible to do this, because the industrial classes whose interests are equally involved are represented in our national and State legislatures by persons having little or no economical interest in common with them. An analysis of the employments of the members of our present Congress shows the following extraordinary facts:

Agriculture, employing 5,922,477 persons, is represented by 26 farmers and planters, or one for every 228,000 persons.

Professional and personal pursuits, in which 2,684,793 persons are engaged, are represented by 228 lawyers, 12 editors, 3 physicians, 2 professors, 2 civil engineers, and 1 hotel proprietor; in all 248 members, or one for every 10,800 persons.

Trade and transportation, employing 1,191,238 persons, are represented by 32 merchants, 11 bankers, 2 railroad managers, and 1 stage proprietor; 46 in all, or one for 26,000 persons.

Manufactures, mechanics, and mining, employing 2,707,421 persons, are represented by 11 manufacturers and 2 miners, or one for 208,200 persons.

The occupations of 33 members is not given, or unknown.

Such a body, however honest and free from the influence of unworthy motives, could hardly be expected to understand and appreciate the needs of the agricultural nor of the other producing classes. Its whole composition is opposed to the idea of representation according to interests involved.

If we examine our State legislatures we find in many instances a very similar state of affairs. Illinois is eminently an agricultural State; yet it has but 8 farmers in its Senate of 51 members, and about 40 farmers in its House of Representatives of 153 members. Although the State contains but $3\frac{1}{4}$ lawyers for every 1,000 of her people, yet the members of this profession comprise by far the most numerous class in her legislature, and 19 of her 21 members of Congress are lawyers.

But while the industrial interests are thus scantily represented or misrepresented in Congress, we are told that some 80 of these

Congressmen are directors and stockholders in national banks; a large number are interested as directors, stockholders, and attorneys in the railway corporations of their own and other States; a smaller number are directly and indirectly interested in protected manufactures. The remainder are members of the legal fraternity, not generally the ablest or best of their class, habituated to taking the fees and arguing the cases of all comers, and not always clearly distinguishing the duties of the representative to his constituency from those of the advocate to his client. On a smaller scale, and with considerable variation and qualification, the same facts appear in our State legislatures.

III. The consequences are what might be anticipated. Legislation is shaped to directly meet the interests of those engaged in legislation, of the classes to which they belong, or by which they are retained. Hence we have, first and worst of all, legislation whereby special privileges are given, or attempted to be given, to a favored few in perpetuity to tax the public. The railroads, which practically are now the highways of the country, are turned over to corporations that claim the rights of public corporations, but refuse to perform their duties. The practical result is, that the whole transportation of the country, but especially freights from the West to the sea-board, and from non-competing points, is compelled to pay a heavy tax over and above the necessary cost of transportation. Every bushel of wheat or corn or other article produced in excess of home consumption has its price depressed far below its proper value, whether it be transported or not. The ten States of the Northwest, producing annually more than a million bushels of grain, find the market value of this grain depressed from 20 to 25 cents per bushel by extortionate charges for transportation. You can see what vast possibilities for ill-gotten gain on the one hand, and enormous loss on the part of the farmers of the West are implied in that statement. Hence it come, that the railroads of the United States, having at the end of 1872 a capital stock of $1,647,844,123, and claiming a total cost of $3,159,423,057, according to Mr. Poor, although generally extravagantly built and often dishonestly managed, paid 5.2 per cent. on their alleged cost, and nearly 10.5 per cent. on their capital stock, while the agricultural wealth of the country was increasing barely 4 per cent. per annum.

Again, take banking corporations, such as are now organized in our national banks, and we again see the result of special privilege conferred by law. The same capital is permitted at once to draw 5 or 6 per cent. coin interest, and 8, 10, or more per cent. currency from private persons.

IV. Legislation and other governmental powers are in like manner used to confer special privileges less objectionable in their character than those I have just mentioned, but often very onerous while they exist. Of this class are "protective" tariffs and patent "rights." In both these cases the representatives of the people attempt to do indirectly what they dare not do directly. It is conceivable that direct bounties to encourage the establishment of manufactures, and pecuniary reward to meritorious inventors, may be sound public policy, but what are we to think of a policy that not only taxes the people for more than the cost of direct subsidies, but lays the foundation of monopolies that extort from the consumer from four to five times the cost of production? Yet such we find to be the cause of the enormous profits on reapers and mowers, sewing-machines, and the high cost to consumers of iron, salt, lumber, etc., compared with the expense of production. Here, especially, in the "ways that are dark" of Congressional lobbies and departmental corridors, does the manufacturing or "royalty" speculator follow his lucrative pursuit, which is to end in the consumer paying $75 for a sewing-machine that is made for $15, and $200 for a reaper that cost $50. This, too, is more than 4 per cent. per annum.

V. Leaving the discussion of direct special privileges granted by legislation, we next elicit the fact that taxation, national and State, is to a great extent levied upon consumption, that is, upon the person, and little upon property, which theoretically, it is conceded, should bear a large portion of the burden; and, furthermore, that where taxation of the wealth of a State has been attempted, it has been the tangible or visible property of the industries, and especially of the farm that has paid the tax in a very unequal degree. Take our national revenues, derived, according to the last report of the Secretary of the Treasury, about

58 per cent. from customs, 35 per cent. from internal revenue, and the remaining 7 per cent. from miscellaneous sources. The 35 per cent. from internal revenue is derived mainly from tobacco and spirits. The 75 per cent. of the population engaged in production pay three quarters of this tax, although they own less than half the wealth of the country. The revenue derived from customs is assignable in nearly the same way; and I feel safe in saying that 70 per cent. of our national taxes are paid by men owning 45 per cent of its wealth; in other words, that the wealth of the producers pays three dollars national taxation for one of the non-producers. Take our State taxation, and we find in most States of the Union that in attempting to assess and tax all kinds of property, tangible and intangible, the result has been a failure to reach the invisible property to a great extent, while the farms, live-stock, shops, and tools of the industrial classes and country districts have been almost exhaustively assessed. In Illinois, where agricultural wealth is a little more than half the total wealth, the agricultural taxation is estimated at from 60 to 75 per cent. of the whole, or from 10 to 25 per cent. more than is just. This is the kind of taxation to which wealth paying 4 per cent. per annum must submit, while the more profitable capital of the country escapes by paying from one third to one half that rate.

VI. The legislation of Congress upon the money or currency question, while baneful to all classes engaged in legitimate trade and manufactures, has been specially injurious to the agricultural class, and only helpful to their worst enemies, the great speculators in railway stocks and their allies. Our currency now has a nominal value of say 90 cents, as compared with the gold dollar; and practically we might expect to buy and sell in that currency at an advance of say 12 per cent. on former prices. But the facts are far otherwise. Corn, wheat, flour, and lard averaged the same or a less price in the New York market during the last five years than they commanded in gold during the five years preceding the war. But coffee and tea have nearly doubled in value, and sugar increased 38 per cent. Taking a wider range of articles, I find that seven articles of agricultural production—mostly exported —wheat, corn, flour, cotton, leather, mess beef, and lard, have increased in value only 5 per cent. over the gold prices of

1855–59, while four important articles of farm consumption— generally imported—coffee, cut nails, sugar, and tea, advanced in price 40 per cent. Thus the tillers of the soil are selling for less and paying far higher prices than at any time during the last twenty years, because our currency is depreciated.

VII. As a combined result of the opportunities afforded by vicious legislation, we find that an unusually large proportion of the population has engaged in trade and other non-productive employments. Speculation and a large element of uncertainty pervades all business. Monopolies founded on the special privi- leges granted by legislation are made more permanent and mis- chievous by private conspiracy. Railroad and other transporta- tion companies combine to fix rates, publishers to fix the price of school-books, and manufacturers to sustain the prices of plows. We have "rings" in legislation, "corners" on 'Change, and "pools" in railway earnings, that threaten to throttle all honest and honorable trade transportation and toil. Such is the condi- tion of our country, under which first of all the agricultural class, but ultimately all the industrial classes, must succumb, unless ca- pable of resistance and reform. All these abuses, discussed in the light of a more intelligent and pronounced Republicanism, tend to build up a new political organization, which, appealing to the first principles of the Declaration, comes forward to abolish the new slavery with which corporate wealth threatens us, and to put out of the way some of the baser forms of class privilege that now infest our Democracy.

The Farmers' Movement in the Western States means, then, first an advance in intelligence and ability on the part of the till- ers of the soil; secondly, a recurrence of one of those periods not uncommon in history of unusual oppression and distress caused by bad government, and resulting in rebellion on the part of the oppressed; and finally, an effort to reverse the unwise legislation that has, in the guise of corporate and other monopolies, created, fostered, and perpetuated a Shylock aristocracy, whose nobility compels no nobleness, but, insatiate, plunders rich and poor with a cruel impartiality.

The result I think will be to carry to yet more logical conclu- sions the principles of our Republican Democracy. It is a part, and an important one, of the general movement among the manual workers of the world. It partakes of the nature of that irrepressible conflict which overthrew one form of oppression, and is as inevitable as the progress of Democracy on the earth. It means that the time draws nearer when the cunning of the hand shall be directed by the brain of the worker—and not by the back of the taskmaster. And *that* means a more equal division of profits—a more pleasant life for the laborer and a simpler and more republican life for those who would thrive by others' toil. Our agricultural colleges and polytechnic schools—the very free schools of our country districts—make it the more certain: for intelligent labor will not submit to the brutal despotism of corporations, and will demand and have its rights. And so far as the agricultural class are concerned, it is not a suddenly taken, inconsiderate action. The farmers of the country, above all others, have given hostages to fortune, not only in wife and children, but in houses and lands, herds and crops. They are by necessity conservative in action, but they have a keen sense of justice and have endured at least so long as endurance is a virtue.

8 FROM   *Charles Francis Adams Offers Judgement on the Granger Movement, 1875*

Charles Francis Adams, grandson of John Quincy Adams, was a frequent contributor to the North American Review *on railroad and other subjects. He became a member of the Massachusetts Board of Railroad Commissioners in 1869 and in the 1880s served as president of the Union Pacific Railroad. He writes from a broad knowledge of railroads and the difficulty of regulating them. This article, no doubt considered too tolerant and sympathetic by many railroad interests, considers the Granger Movement only as a railroad reform movement. Adams simply ignores the National Grange as an organization dedicated to the educational, social and economic betterment of its members. For a biography of Adams, see E .C. Kirkland,* Charles Francis Adams . . . *(Harvard, 1965).*

SOURCE. *North American Review* 120:394–424. (April 1875).

ARTICLE V   THE GRANGER MOVEMENT.

The great Know-Nothing movement, so called, which swept over the United States about twenty years ago, apparently originated without cause, raged subject to no law, and finally subsided, having produced no permanent results. It was a species of popular squall preceding the long, violent tempest of the Rebellion. In this respect it furnished a striking exception to the general principles which mark the rise and development of widespread popular agitations. They seldom originate without cause; and, in spite of blunders and mismanagement, rarely pass away without having contributed something worth having to that general result which makes up the conditions under which we live. This is especially true of that Granger movement, which, during the last four years, has played a most prominent part in the politics of certain of the Northwestern States, and resembled the Know-Nothing movement only in its more prominent and least creditable features. It is quite apparent, however, that the time has now come when the Granger can be looked upon as a phenomenon of the past, and treated in a spirit of critical justice. Hitherto this has not been done. In the West the Granger has been "inside politics," and the politicians and editors of rival factions have vied with each other in flattering his vanity, extenuating his shortcomings, and excusing his misdeeds. In the East, on the other hand, the public mind has been mainly impressed by certain striking episodes in his movement, which, naturally, were almost always those reflecting the last credit upon it: of this character was the defeat at the polls of Chief Justice Lawrence of Illinois, for having presumed to decide a constitutional issue which arose before him as a judge, on principles of law rather than in obedience to a popular demand: or, again, the "Potter Law," so called, of Wisconsin, which seemed designed to operate as a practical confiscation of many millions of foreign capital invested in the public improvements of that State; or, finally, the Illinois Railroad Law, which was ingeniously framed so as to make those who were to use the railroads of Illinois the final arbiters as to what it was reasonable they should pay for such use. Such episodes as these have led the people of the East to regard the

Granger movement as one of those causeless, unjustifiable, and outrageous manifestations, nearly allied to agrarianism, which attempt to perpetrate under the forms of law the most wanton assaults upon property. It is not necessary to have a very exalted opinion either of the Granger movement or of the Granger type of politicians and thinkers to assert that the conclusion last stated does it and them scant justice. The simple truth is that the Granger excitement was not causeless, and that, in spite of the blunders which marked its career; it has done a great deal of very good work.

To appreciate this or any other popular movement, it is necessary to have a somewhat definite conception of its surroundings, —of the various conditions, social, political, economical, and geographical, among which it manifests itself. So far as they are necessary to an intelligent judgment, these may, in the present case, be very briefly stated, and they are not without their interest. In the first place we have a young agricultural community, which has grown up with unprecedented rapidity in the very centre of the continent, many hundreds of miles away from those great centres of human industry to which it must look for its markets and sources of supply.

. . . within the last ten years, the public mind of the West has dwelt upon railroads in two wholly different moods. During the first five years,—those between 1865 and 1870,—the fact of geographical isolation was seen clearly and felt strongly. The insufficiency of Western capital and the pressing need of new channels of communication with the East assumed an ever-increasing prominence, and thus railroads, more railroads, were the constant longing of the Western man and the unceasing burden of his speech. He found his way—loaded with maps and plans and prospectuses, and stock and bonds and land grants—into every money market of the world. Railroads could not, he thought, be purchased at too high a price; no inducements were too large with which to tempt foreign capital. The Western imagination was thoroughly fired; and so, at last, was Eastern cupidity:—and the result was that disastrous railroad mania which culminated in the panic of 1873. Through a period of five years capital flowed to the West in an apparently inexhaustible stream, and under its influence railroads were constructed as if by magic. The best and most preposterous lines were equally built; competing line was run upon competing line between the great centres; while other lines were laid out from points where no one lived to points where no one wanted to go. In the Western States and Territories alone, during the five years which preceded the outbreak of the Granger movement, not less than five hundred million dollars of actual wealth were invested in the construction of railroads. Accordingly the number of miles of road in operation in Illinois is stated to have risen from 3,191 in 1866 to 5,904 in 1871; those in Wisconsin, from 1,036 to 1,725; those in Iowa, from 1,283 to 3,160; those in Minnesota, from 482 to 1,612; those in Nebraska, from 473 to 943; while Kansas during the same period ran up from 494 miles to 1,760. These are the six essentially Granger States, and in them alone the increase of railroad mileage in the six years between 1867 and 1873 was from 6,992 to 17,645, or no less than 254 per cent.

Thus about the year 1870 the Western mind began to appreciate the fact that its railroad system was secured. Not unnaturally it now began also to count its cost, and to realize that, though the West had the use of this magnificent railroad development. and could not be deprived of it, yet she did not own it, and was, moreover, bound to pay for its use according to the bonds given anterior to its construction. In other words, exactly what might have been anticipated now began to appear. In her over-eagerness the West had made an improvident bargain; she had given for her longed-for railroads all that she had, all that any one asked; and now she had them, and began to shrewdly suspect that her bargain had after all been somewhat of the hardest. It was, indeed, a case of absentee ownership, with all that those words imply; and when that is said one great cause of the near-impending trouble is disclosed.

. . . grievances originated in two causes, and to one or the other of these two causes could generally be traced nearly all that hostility which gave the Granger movement such power and public sympathy as it had; these were, in the first place, Competition, and, in the next place, Bad Manners.

The people of the West eagerly invited foreigners to build railroads for them; and it was not until after their system was practically constructed that they were made, through bitter experience, to realize that competition between its members was producing results neither such as had been anticipated nor such as were altogether satisfactory. They found, in a word, that while the result of ordinary competition was to reduce and to equalize prices, the result of railroad competition was to produce local inequalities and to arbitrarily raise and depress prices. The railroads of the West had been built a great deal too rapidly, and the business of the country could not support them; those immediately in charge of them were under a heavy and unceasing pressure to earn money, and they earned it wherever and however they could,— where it was in their power to earn it through exactions, they exacted; where they were forced to compete for it, they competed. There resulted a system of inequitable local discriminations which might not unfairly be described as intolerable. At one point several roads would converge, and the business or travel from that point would be furiously fought over and done for almost literally nothing; while other points but a few miles away would be charged every dollar that their business could be made to pay without driving it off the railroad and back into the highway. Where goods were started from the same point to different stations upon the line of the same road, those forwarding them discovered to their cost that the tariffs resembled nothing so much as an undulating line,—for a distance of twenty miles, more would have to be paid than for one of forty miles; and not infrequently a consignee would see goods habitually carried by his door to some point miles farther on, in order that the company might charge him the local rates for bringing them back. Those living between competing points were rigidly excluded from the benefits of competition. To such an outrageous extent was this carried, that it became the common practice where an entire car-load of merchandise was paid through to a competing point to make a large extra charge for *not* hauling it to that point, but leaving it at its ultimate destination, perhaps a hundred or two miles short of it. Competition led also to favoritism of the grossest description,—men or business firms whose dealings were large could command their own terms as compared with those whose dealings were small. The most annoying and injurious inequalities were thus spread all over the land. Every local settlement and every secluded farmer saw other settlements and other farmers more fortunately placed whose very prosperity seemed to make their own ruin a mere question of time. Man to man, or place to place, they might compete; but where the weight of the railroad was flung into one of the scales, it was strange if the other did not kick the beam.

Of course, even under the most favorable circumstances, such a condition of affairs could not be perpetuated. In this case, however, it was aggravated by a system of gross jobbery and corruption which, before the storm burst, seemed fairly to have honeycombed the whole railroad system of the West. It began high up in the wretched machinery of the construction company, with all its thimble-rig contrivances for transferring assets from the treasury of a corporation to the pockets of a ring. Thence it spread downward through the whole system of supplies and contracts and rolling stock companies, until it might not unfairly be said that everything had its price. The whole story is, however, told in these two words, *Absentee Ownership;*—while the Western patron was plundered, the Eastern proprietor was robbed. Under these circumstances the continuance of the system was made even shorter than it otherwise need have been by the other cause of grievance which has been referred to,—Bad Manners.

This is a vastly more important matter to railroad corporations, not only in the West, but all over the country, than those owning or managing them appear to be aware of. In New England the conditions of affairs is bad enough; and more than one important corporation has experienced great injury, or finds itself leading an existence of perpetual warfare and turmoil, solely through the inability of some prominent and, perhaps, otherwise valuable official to demean himself with consideration towards his brother man. What in this respect is seen here in the East is absolutely nothing to what prevails in the West. Taken as a class, the manners of the employees of the Western railroad system are probably the worst and most offensive to be found in the civilized world.

So now the railroads were no longer the pioneers of dawning civilization or the harbingers of an increased prosperity; they were the mere tools of extortion in the hands of the capitalists, —the money-changers of the East,—marauders, banditti, usurers, public enemies.

The legislative struggle was, however, soon over. Laws were passed and went into effect under which entire tariffs of charges were imposed upon the railroad corporations. These, of course, the corporations resisted in the courts, not always in the most judicious manner, upon the ground that under the Constitution of the United States the Legislatures had no authority to practically confiscate private property by decreeing that the public might enjoy it on paying therefor an inadequate compensation, or no compensation at all if it so saw fit. On this question more will be said presently. One of the least creditable features of the Western character now, however, began at once to reveal itself, —an extreme restiveness under the restraints of law. This is a not unnatural remnant of the old frontier life, and one which is rapidly passing away; but in 1873 it is none the less an historical fact that a majority of the voters of what is beyond question the leading State of the Northwest could not and would not understand how any court of law should presume to set aside as illegal a duly authenticated expression of the popular will.

Apart from questions of discrimination and those relating to the local management of their railroads, the Granger agitation was based upon two hypotheses, neither of which will bear a close examination. The first of these was that the West was paying to the corporations owning its railroads an inordinate profit on the work of transporting its products to the seaboard; and, second, that these profits were made necessary to pay dividends upon an exorbitant and fictitious cost of the roads. The first proposition generally took the form of a statement, more or less correct, as to the proportion borne between the value of a bushel of corn in the West and the cost of transporting it to a market at New York; in general terms it would be said that of five bushels of corn, four were taken by the railroads as the price for carrying the other one to a market. This may perfectly well be true, and yet signify absolutely nothing. Before finding fault with what is paid for a service, it is desirable to know what the service is:— what, in this case, was the distance of the market? As a matter of course there must be a point somewhere to which transportation would cost more than the entire worth of the article transported, especially if that article should chance to be heavy and of small intrinsic value. The transportation-tables which have long been in use show that upon a common earth road corn of the ordinary value can be carried only 165 miles before its whole money-worth will be consumed in the cost of its transportation. By rail, however, it can be carried some 1,650 miles. Because it can be carried this much greater distance, it is manifestly absurd to claim that there is no limit to which it may *not* be carried before that result should be arrived at; the only question is whether it is carried a proper and reasonable distance. On this point the Granger authorities have never met the statistics presented by the railroad corporations, and, indeed, it is not easy to see how these could be met. Under the stress of competition over great distances, it would be found that, as respects this class of business, the usual and obvious result had been arrived at,—that peculiar description of merchandise is carried at less than cost, and something else has to make good the loss incurred upon it.

Neither are the Granger authorities more fortunate in their second proposition, that the extortionate charges of the railroad corporations are made necessary in order that dividends or interest may be paid upon an excessive and wholly fictitious cost of the railroad system.

Here, then, are the five great Granger States,—those in which the cry against the unfeeling extortions and the inordinate profits of the railroad "monopolists," "Shylocks," "marauders," banditti," "feudal barons," and "tryants" has rung out loudest and longest; yet what are the conclusions to be drawn from the returns? Those returns include 25,000 miles of railroad. It may be objected that a portion, and no inconsiderable portion, of this amount enters more than once into the total. That fact, however, does not affect the results, as the securities, the discounts, and the earnings all enter into the computations in the same degree. These 25,000 miles of road are represented by $1,130,000,000 of paper securities, upon which the net earnings of the system equal, not eight per cent, but just half of that amount, four per cent. Casting out of this great aggregate thirty-six per cent of fictitious capital, reducing it at once by $16,000 per mile and by a grand total of $400,000,000 to a dry, waterless basis, we find that the aggregate of the net earnings represent an annual return of just 6.5 per cent on the investment. This certainly is neither usurious nor oppressive.

It is difficult, therefore, from anything which can be found in the statistics upon the subject to avoid the conclusion that, though the States of the Northwest got their railroad systems at a very high nominal cost, yet that they paid for them largely in the most worthless of paper securities; and that the real cost, after the slow process of liquidation has worked itself out, will prove to be not only actually reasonable, but even considerably less than it would have been had the investment been guaranteed by the State governments. It would also seem that the people of those States have no just cause of complaint as respects the cost of moving their products to the seaboard. There, at least, competition has produced every result which even the most sanguine could have anticipated from it. What, then, has been the real underlying, hidden cause of this widespread agitation?—was there any, or was it, after all, a mere restless surface movement? The real cause of complaint, the true source of the evils under which they suffer, has as yet received but little mention among Western men; in fact, the subject is one the discussion of which they instinctively avoid, for there are no votes in ugly truths. Though the source of all their woes is not apparent on the surface, it may be described in very few words,—*they have gone too far West.* For this they are themselves chiefly, though not wholly, responsible. The West has ever proved itself the steady, reliable ally of that wretched land-grant and subsidy policy which did so much to stimulate the mania for railroad construction. For years the ruling idea of the Western mind has been the bringing of remote acres, and ever acres more remote, under cultivation. There was thought to be some occult virtue in expediting this process,—a service to God and one's country. Every artificial appliance and inducement was thus set to work to force population out in advance of the steady and healthy growth of civilization into regions beyond the reach of the world's centres and outside the pale of social influence. It was this hurtful forcing process which brought about that condition of affairs which had to culminate in the Granger movement, and in the organized assault on property in railroads. The people were paying the penalty of too rapid growth,—paying it just as much as any boy or girl must pay it who is so unfortunate as to outgrow strength and clothes at once.

The result brought about by the unnatural diffusion of population, so far as the agricultural interests of the West were concerned, was exactly what any thinking and observing man should have anticipated,—over-production at remote points. This difficulty no increased cheapness of transportation can alleviate; it can only transfer the locality of the difficulty to a point somewhat more remote. The darling vision of the Granger's dreams, the Utopia of his waking fancies, and the constant theme of his noisy rhetoric, is a double-track, steel-rail, government-built, exclusively freight railroad from every farmer's barn-door straight to the city of New York. Paradoxical as it may seem, there is not the slightest room for doubt that even the full realization of this fanciful impossibility would not at all benefit the individual farmer of the West. It would fail to benefit him for a very simple and obvious reason. The difficulty he is now laboring under is over-production; the West grows more of the fruits of the soil than the world will consume at present prices. Meanwhile the area from which production is possible is not only not fully occupied, but is for all practical purposes unlimited. A reduction of the present cost of carriage, therefore, only serves by so much to extend the area from which the supply can be drawn, brings so many additional acres and so many more farmers into the field of competition. The whole benefit of the reduction inures, therefore, not to the producer, but to the consumer. The new-comers glut the market before it can be relieved. Any reduction in the cost of the carriage of agricultural products is, therefore, of enormous importance to us on the Atlantic seaboard, and of yet more importance to the swarming population of the British isles,—to the competing agriculturists of Eastern Europe it involves also most serious consequences,—but to the farmers of the West, as a class, it amounts to nothing more than one additional step in continuance of that same progress which has steadily been going on for over thirty years, and which they now claim has brought them to their present hard and desperate pass. Ever since 1830 the cost of transportation has been growing cheaper and cheaper, until it has now touched points which would once have been considered incredible; yet the standing complaint of the farmer is still that the cost of carriage consumes the whole value of his product; just as much so to-day, when the limit of its carriage is sixteen hundred miles, as fifty years ago when it was but one hundred and sixty miles.

The Granger movement touches, then, the real cause of the evil under which the West is suffering only so far as it tends to supplement the disasters of the recent financial crisis and put a complete stop to all further immediate railroad construction. In this way it may help to hold in check the existing tendency of population to diffuse itself prematurely, and restore the country to a healthy, because measured process of development. This, however, is a result which its leaders and would-be philosophers have not contemplated, and which partakes, indeed, somewhat of a boomerang character. They may, however, yet learn that what we need is not always that which is pleasant to get, and that we sometimes build more wisely than we know. Meanwhile the recent halting and confused action of the legislatures of the Northwest makes it apparent enough that the Granger flood is reached, and that the ebb will soon follow,—the movement is now obviously losing its strength. Though it accomplished little that it intended, it has yet, unconsciously to itself and through that rough process of attrition by which most results that are valuable are brought about, removed or greatly modified those more superficial grievances which gave it its only popular strength. It has placed many preposterous laws on the statute-books of the West, which will probably long remain there, undisturbed memorials of legislative incapacity, and about as formidable as those ancient blunderbusses which sometimes in old-fashioned houses ornament the kitchen wall. Undoubtedly it has seriously impaired the credit of those States more especially identified with it, and notably that of Illinois and of Wisconsin. For this, of course, they will have to pay dearly;—higher interest and more binding guaranties will unquestionably be exacted of them, and, what is more, they will have to give them. Habitual borrowers cannot afford to play tricks with their credit, and it will be very long indeed before either the defeat of Judge Lawrence or the provisions of the "Potter Law" are forgotten in Wall Street or on the Royal Exchange.

In some respects the results produced by the movement have been most beneficial. The corporations owning the railroads have been made to realize that those roads were built for the West, and that, to be operated successfully, they must be operated in sympathy with the people of the West. The whole system of discriminations and local extortions has received a much-needed investigation, the results of which cannot but mitigate or wholly remove its more abominable features; finally, certain great principles of justice and equality, heretofore too much ignored, have been driven by the sheer force of discussion, backed by a rising public opinion, into the very essence of the railroad policy. All this is much gained. The burnt child fears the fire, and the Granger States may rest assured that, through an indefinite future, the offensive spirit of absentee ownership will be far less perceptible in the management of their railroads than it was before and during the great railroad mania. Finally, East and West, the good which has resulted and yet will result from the Granger movement will be found greatly to predominate over the evil; what is more, the good will survive, while the evil will pass away.

C. F. Adams, Jr.

9 FROM *The Nation Speaks About the Grangers, 1876*

The Nation, *edited by E. L. Godkin from 1965 to 1900, was described by James Bryce as the "best weekly not only in America but in the world." Yet this journal showed little understanding and almost no sympathy for the troubled farmers either during the period of the 1870s or later, when the prairie fires of farmer protest burned in the agricultural Alliances and Wheels. Like Charles Francis Adams, Godkin and the* Nation *seemed largely unaware of the National Grange. See Rollo Ogden,* Life *and Letters of Edwin Lawrence Godkin (New York, N.Y., 1907) and William A. Russ, Jr. "Godkin Looks at Western Agrarianism: A Case Study," in* Agricultural History, *19:233–242.*

THE GRANGER COLLAPSE

The inaugural address by the new Governor of Wisconsin brings out strikingly the great change of feeling on the subject of Granger legislation which has come over the Northwestern

SOURCE. *The Nation*, 22:57–58 (January 27, 1876).

States within the past year. The whole question of the validity of the laws passed by Wisconsin and other States is now before the Supreme Court, and will soon be decided by that tribunal finally and without appeal. But whichever way the decision may go as to the principle of the laws, the question whether they have worked well—whether they have benefited the States which have adopted them—is a practical question which must be decided by the Grangers' own experience; and it is this question on which the farmers have now pretty much made up their minds. In Wisconsin, it seems from the Governor's address, experience shows that they have produced the following effects: Railroad construction, even in those parts of the State which most need it, has come to a dead stand-still; no company has paid dividends on its stock for the past two years, and during the past year only four companies have paid interest on their bonds. If this cessation of dividends and interest meant a diversion of the money into the pockets of the oppressed farmer, it might be very well; but unfortunately the farmer is a borrower of money, and the farming States of the West depend for their prosperity upon the stability of their credit; and the Governor declares that while the right to pass Granger laws is one "of vital importance," it cannot at the same time be denied that "the existing laws" have, "either justly or unjustly," "impaired the credit of the State, and of *its individual citizens,* in the commercial and financial centres of the world." With "immense resources undeveloped," and a "consequent need of capital from sources where it is in excess," the people "find capital repelled" by these laws. This the Governer thinks is not strange, inasmuch as everybody outside the State can see, 1st, that capital has been invested in Wisconsin railroads; 2d, that they are in daily use by the people of the State; 3d, that this use pays nothing to the owners; 4th, that the owners are compelled by law to permit the use, and "deprived by law of the right to say what they shall receive for it." He therefore recommends the repeal of the Granger laws, and the substitution for them of provisions against extortion and unjust discrimination, as well as the establishment of "maximum rates" "not greater than those fixed by the companies when they made their own tariffs"—the latter suggestion being, of course, a convenient device for covering the Granger retreat. The laws will, in all probability, be repealed this winter, and as the Grangers ceased some time since to influence general politics, we may fairly be considered to have arrived at the end of the agitation.

Our readers will, we believe, bear us out in saying that we have always predicted this result from the time when the new movement first showed its real character. We did so not because a movement against railroad abuses was at all distasteful to us, but because we have never believed that a movement would succeed in America which was really directed, not against abuses, but against the rights of property. That a particular class in the community would, under cover of a vague cry about oppression and wrongs, the exact nature of which they were never able to state clearly, be long allowed to masquerade as reformers for the purpose of making a living at their neighbors' expense, we have always had too high an opinion of American honesty and common sense to suppose. When the Grangers had once proclaimed that their object was to "fix rates," or, in other words, to declare by law what proportion of the market value of services they themselves should pay, and that they would not be bound by the terms of their contracts, it was perfectly clear that the Granger movement was rank communism, and its success in this country was against all reason and experience.

At the time when the Granger fury was at its height, those who endeavored to expose its fallacies received a great deal of advice from persons who felt it their duty to "labor" with all doubters, and to show them that a great movement was really on foot and was going to "revolutionize the country." These instructors of the public now maintain a judicious silence, and apparently have entirely forgotten the fearful perils which "watered stock" and railroad "combinations" and "pools" and "consolidation" and "Tom" Scott and Vanderbilt were going to bring in their train. The State, they assured us, must either absorb the railroads or the railroads would absorb the State. Unless we all became Grangers, this was the inevitable end to which we were tending. We trust that the collapse of the Granger movement will be a warning to all who shared these gloomy apprehensions, and who tried to persuade themselves and their neighbors that the way to remedy them was to plunder the railroad companies. When another attempt is made to "revolutionize the country," or to break up the old parties, we trust they will ask themselves, not how many votes the new movement probably numbers, but whether it has a solid reason for existence; whether the arguments of its supporters are true, or mere hollow imitations of reasoning; whether the grievances against which it is directed really exist, and whether, if they do, the movement will remedy or merely increase them; and, finally, whether the character of the men who advocate the new ideas is solid or flimsy; whether they are men who can defend the stand they have taken by their intelligence and reputations, or fury and nothing more. If they cannot satisfy themselves on these points, we strongly advise them to let the movement rage, and to believe that honesty and good sense will outlive it, no matter how many States it may carry.

PART THREE

Wheels, Alliances, and Populism

The Alliance Movement of the 1880s and the rise of Populism had much in common with the National Grange of the Patrons of Husbandry and the related Granger Movement. The National Grange was dedicated to the improvement of the farmer's life. Its leaders sought to eschew partisan politics. State and local leaders, however, were active in sponsoring and supporting liberal, reform, antimonoply, and other independent parties. After the collapse of the reform movement in the 1870s, the National Grange continued as an organization of farmers and others interested in agricultural pursuits. In contrast, the Alliance organizations, did not survive the political campaigns of the 1890s.

A host of farmer Alliances, Wheels, Unions, Clubs, and other orders rose in the 1880s. They were most numerous and active in the cotton South and in the recently settled wheatlands west of the Missouri River. The initial programs of education and self-help through the organization of cooperative grain elevators, cotton gins, cotton marketing associations, and cooperative stores did not bring relief. Various orders moved to form state, regional, and national associations. Among these, the National Farmers' Alliance and Industrial Union of the South—it claimed 2 million members by 1890, more than the number of farms in the South—and the National Farmers' Alliance of the upper Mississippi valley were the most active.

Although the Alliances and other farmer orders declared initially that they would avoid partisan politics, the growing list of grievances against bankers and creditors, the money and credit system, the railroads and other corporations, the commission merchants, and the unresponsiveness of state and national government, caused them, by 1890, to support or actually to conduct political campaigns. In the South they worked inside the Democratic party; in the North they created or helped create numerous independent parties under various names. The Alliance groups and their allies enjoyed considerable success in electing state legislators, some governors, and members of Congress.

This success marks the beginning of the absorption of the Alliance Movement into a third party that offered a broad reform program. But the party was also characterized by an increasing obsession on the part of many of the leaders and supporters with cheap money, particularly free coinage of silver, as being essential to almost all of their reforms. The new party, The People's party, usually referred to as Populists, enjoyed some success in the elections of 1892 and 1894. But in 1896 the Democratic party adopted the free coinage of silver plank and virtually absorbed the People's party. It did not survive the defeat of 1896 with any strength, but neither of the major parties was to be quite the same thereafter.

Meanwhile, the Alliance Movement subsided and the state and local organizations eroded and disappeared, some of their members enlisting in the Grange or in the newer farmer organizations that would rise in the 20th century. Carl C. Taylor has published a melancholy account of a meeting he attended in 1928 of one of the vestigial farmer alliance groups that still survived in North Carolina [Carl C. Taylor, *The Farmers' Movement*, (New York, 1953) pp. 325–326].

As is true of the Granger Movement, full-length histories of the Alliance Movement appeared promptly. W. Scott Morgan, *History of the Wheel and Alliance, and the Impending Revolution,* was the first to be published (Fort Scott, Kansas, 1889). It was followed by N. A. Dunning, *Farmers' Alliance History and Agricultural Digest* (Washington, 1891) and a spate of other books. John D. Hicks, *The Populists Revolt: A History of the Farmers Alliance and the People's Party* (Minnesota, 1931) was the first full-length scholarly study to appear.

10 FROM *W. Scott Morgan*
The Changing Purposes of the Farmers' Alliances

The modest declarations of the organizers of the Agricultural Wheel in Arkansas and the Farmers Alliance in Texas, like those of earlier farmers clubs and other farmer orders of the 1880s, showed that their members were seeking to improve themselves, their craft, and their society through education, through the collection and dissemination of agricultural information, and through unified action in buying and selling what they produced and what they needed. These actions, even when successful, did not bring relief. With the organization of state and regional alliances, the original, simple purposes were almost lost in the lengthening list of grievances that provided a basis for the successive demands. The demands agreed to at St. Louis and Ocala exhibit most of planks that appear in the Populists platform in 1892.

THE BIRTH AND EARLY PURPOSES
OF THE AGRICULTURAL WHEEL

The Wheel was born on the 15th of February, 1882, in an old log school house, eight miles southwest of the town of Des Arc, in Prairie county, Arkansas; its parents, as stated before, were monopoly and oppression.

THE MAIDEN WHEEL SPEECH

"GENTLEMEN:—As we have assembled to-night for the purpose of devising some plan by which to extricate ourselves from the grasp of monopoly, we believe it our duty first to take a view

SOURCE. W. Scott Morgan. *History of the Wheel and Alliance and the Impending Revolution* (Fort Scott, Kansas, 1889) , pp. 62–65.

of the adverse circumstances which surround us, to-wit: Drouth, poverty and oppression by organizations whose avowed object is the reduction of laborers to financial and political slavery. I, gentlemen, for one, feel deeply interested in this little organization, whatever it may prove to be. We see, we know, that the products of our labor is being wrested from our hands at astonishingly low prices, and afterwards sold at prices so much greater as to seem almost incredible; yea, double price. We, who produced it by the sweat of our faces and that of our beloved wives and children, are publicly robbed in order that a few heartless and soulless middlemen may be made rich; who, after having acquired wealth in this thieving manner, have the audacity to turn up their sinful noses at their innocent victims dressed in plainer clothes. We have borne these outrages so patiently and so long that the world is fast coming to the conclusion that all is fair in love, war and trade; but it is a falsehood, and the devil is at the bottom of all such conclusions. Not one act of our lives is right unless it is in accordance with the rule that works both ways and toward heaven, 'Do unto others as you would have them do unto you.' Even young men who have been raised on a farm begin to think that swindling behind the counter is more genteel than wearily plodding behind a No. 8 Avery plow. Why is this? The answer is because others get more of the profit of our labor than we do, and even children know it. Were our occupation respected, and this accursed system of monopoly crushed, we would be enabled to beautify our homes and make our sons and daughters love them. Gentlemen, I have not time to-night to give you the many reasons why it becomes our indispensable duty to organize, and that speedily; but I am in favor, 'sink or swim, live or die, survive or perish,' of launching our little bark Union on the rolling billows of opposition, and of lashing ourselves each to his post, until by the help of Israel's God we have ridden safely into the harbor of peace, liberty and prosperity."

ORIGINAL CONSTITUTION

1. This organization shall be known as the Wattensas Farmer's Club.

2. Its objects shall be the improvement of its members in the

theory and practice of agriculture and the dissemination of knowledge relative to rural and farming affairs.

3. The members shall consist of such persons as will sign the constitution and by-laws and who are engaged in farming.

4. Its officers shall consist of one president, two vice-presidents, secretary, chaplain and treasurer, who shall jointly constitute the executive committee; also two sentinels, and shall be elected annually.

5. Its meetings shall be held on the first and third Saturday nights in each month at McBee's school house.

Declaration of Purposes[1] [2]

PROFOUNDLY impressed that we as the Farmers' Alliance, united by the strong and faithful ties of financial and home interest, should set forth our declaration of intentions, we therefore *Resolve:*

1. To labor for the Alliance and its purposes, assured that a faithful observance of the following principles will insure our mental, moral, and financial improvement.

2. To endorse the motto, "In things essential, Unity, and in all things Charity."

3. To develop a better state, mentally, morally, socially, and financially.

4. To create a better understanding for sustaining our civil officers in maintaining law and order.

5. To constantly strive to secure entire harmony and good will among all mankind and brotherly love among ourselves.

6. To suppress personal, local, sectional, and national prejudices, all unhealthy rivalry and all selfish ambition.

DEMANDS OF THE FARMERS' ALLIANCE OF TEXAS, 1886[3]

We, the delegates to the Grand State Farmers' Alliance of Texas, in convention assembled at Cleburne, Johnson County,

[1] Dunning, *Farmers' Alliance History,* p. 28.
[2] Adopted at Friendship, Wise County, Texas, 1880.
[3] Dunning, *Op. Cit.,* pp. 41–43.

Texas, A.D. 1886, do hereby recommend and demand of our State and national governments, according as the same shall come under the jurisdiction of the one or the other, such legislation as shall secure to our people freedom from the onerous and shameful abuses that the industrial classes are now suffering at the hands of arrogant capitalists and powerful corporations.

We demand:

1. The recognition by incorporation of trade-unions, co-operative stores, and such other associations as may be organized by the industrial classes to improve their financial condition, or to promote their general welfare.

2. We demand that all public school land be held in small bodies, not exceeding 320 acres to each purchaser, for actual settlement, on easy terms of payment.

3. That large bodies of land held by private individuals or corporations, for speculative purposes, shall be rendered for taxation at such rates as they are offered to purchasers on credit of one, two, or three years, in bodies of 160 acres or less.

4. That measures be taken to prevent aliens from acquiring title to land in the United States of America, and to force titles already acquired by aliens, to be relinquished by sale to actual settlers and citizens of the United States.

5. That the law-making powers take early action upon such measures as shall effectually prevent the dealing in futures of all agricultural products, prescribing such procedure in trial as shall secure prompt conviction, and imposing such penalties as shall secure the most perfect compliance with the law.

6. That all lands forfeited by railroads or other corporations, immediately revert to the government and be declared open for purchase by actual settlers, on the same terms as other public or school lands.

7. We demand that fences be removed, by force if necessary, from public or school lands unlawfully fenced by cattle companies, syndicates, or any other form or name of corporation.

8. We demand that the statutes of the State of Texas be rigidly enforced by the Attorney-General, to compel corporations to pay the taxes due the State and counties.

9. That railroad property shall be assessed at the full nominal value of the stock on which the railroad seeks to declare a dividend.

10. We demand the rapid extinguishment of the public debt of the United States, by operating the mints to their fullest capacity in coining silver and gold, and the tendering of the same without discrimination to the public creditors of the nation, according to contract.

11. We demand the substitution of legal tender treasury notes for the issue of the national banks; that the Congress of the United States regulate the amount of such issue, by giving to the country a per capita circulation that shall increase as the population and business interests of the country expand.

12. We demand the establishment of a national bureau of labor statistics, that we may arrive at a correct knowledge of the educational, moral, and financial condition of the laboring masses of our citizens. And further, that the commissioner of the bureau be a cabinet officer of the United States.

13. We demand the enactment of laws to compel corporations to pay their employees according to contract, in lawful money, for their services, and the giving to mechanics and laborers a first lien upon the product of their labor to the full extent of their wages.

14. We demand the passage of an interstate commerce law, that shall secure the same rates of freight to all persons for the same kind of commodities, according to distance of haul, without regard to amount of shipment. To prevent the granting of rebates; to prevent pooling freights to shut off competition; and to secure to the people the benefit of railroad transportation at reasonable cost.

15. We demand that all convicts shall be confined within the prison walls, and the contract system be abolished.

16. We recommend a call for a national labor conference, to which all labor organizations shall be invited to send representative men, to discuss such measures as may be of interest to the laboring classes.

17. That the president of the State Alliance be, and he is hereby, directed to appoint a committee of three to press these de-

mands upon the attention of the legislators of the State and nation, and report progress at the next meeting of the State Alliance. And further, that newspapers be furnished copies of these demands for publication.

CONSTITUTION (1887) [4]

Declaration of Purposes

Profoundly impressed that we, the farmers of America, who are united by the strong and faithful ties of financial and home interests, should, when organized into an association, set forth our declaration of intentions, we therefore resolve:

1. To labor for the education of the agricultural classes in the science of economic government, in a strictly non-partisan spirit, and to bring about a more perfect union of said classes.

2. That we demand equal rights to all and special favors to none.

3. That we return to the old principle of letting the office seek the man, instead of the man seeking the office.

4. To indorse the motto, "In things essential unity, and in all things charity."

5. To develop a better state mentally, morally, socially, and financially.

6. To create a better understanding for sustaining our civil officers in maintaining law and order.

7. To constantly strive to secure entire harmony and good will to all mankind, and brotherly love among ourselves.

8. To suppress personal, local, sectional, and national prejudices, all unhealthful rivalry, and all selfish ambition.

9. The brightest jewels which it garners are the tears of widows and orphans, and its imperative commands are to visit the homes where lacerated hearts are bleeding; to assuage the sufferings of a brother or sister; bury the dead; care for the widows and educate the orphans; to exercise charity towards offenders; to construe words and deeds in their most favorable light, granting honesty of purpose and good intentions to others; and to pro-

[4] Dunning, *Op. Cit.*, pp. 58–59.

tect the National Farmers' Alliance and Co-operative Union until death. Its laws are reason and equity; its cardinal doctrines inspire purity of thought and life; its intention is, "Peace on earth and good will to man."

THE ST. LOUIS DEMANDS, 1889[5]

Agreement made this day by and between the undersigned committee representing the National Farmers' Alliance and Industrial Union on the one part, and the undersigned committee representing the Knights of Labor on the other part, witnesseth: The undersigned committee representing Knights of Labor, having read the demands of the National Farmers' Alliance and Industrial Union, which are embodied in this agreement, hereby indorse the same on behalf of the Knights of Labor, and for the purpose of giving practical effect to the demands herein set forth, the legislative committees of both organizations will act in concert before Congress for the purpose of securing the enactment of laws in harmony with the demands mutually agreed.

And it is further agreed, in order to carry out these objects, we will support for office only such men as can be depended upon to enact these principles in statute law, uninfluenced by party caucus.

The demands hereinbefore referred to are as follows:

1. That we demand the abolition of national banks, and the substitution of legal tender treasury notes in lieu of national bank notes, issued in sufficient volume to do the business of the country on a cash system; regulating the amount needed on a per capita basis, as the business interests of the country expand; and that all money issued by the government shall be legal tender in payment of all debts, both public and private.

2. That we demand that Congress shall pass such laws as shall effectually prevent the dealing in futures of all agricultural and mechanical productions; preserving a stringent system of procedure in trials as shall secure the prompt conviction, and imposing such penalties as shall secure the most perfect compliance with the law.

[5] Dunning, *Op. Cit.*, pp. 122–123.

3. That we demand the free and unlimited coinage of silver.

4. That we demand the passage of laws prohibiting the alien ownership of land, and that Congress take early steps to devise some plan to obtain all lands now owned by aliens and foreign syndicates; and that all lands now held by railroad and other corporations, in excess of such as is actually used and needed by them, be reclaimed by the government and held for actual settlers only.

5. Believing in the doctrine of "Equal rights to all and special privileges to none," we demand that taxation, national or State, shall not be used to build up one interest or class at the expense of another.

We believe that the money of the country should be kept as much as possible in the hands of the people, and hence we demand that all revenues, national, State, or county, shall be limited to the necessary expenses of the government, economically and honestly administered.

6. That Congress issue a sufficient amount of fractional paper currency to facilitate exchange through the medium of the United States mail.

7. We demand that the means of communication and transportation shall be owned by and operated in the interest of the people, as is the United States postal system.

For the better protection of the interests of the two organizations, it is mutually agreed that such seals or emblems as the National Farmers' Alliance and Industrial Union of America may adopt, will be recognized and protected in transit or otherwise by the Knights of Labor, and that all seals and labels of the Knights of Labor will in like manner be recognized by the members of the National Farmers' Alliance and Industrial Union of America.

THE OCALA DEMANDS, DECEMBER 1890[6]

Report of the Committee on Demands

Section 1. We demand the abolition of national banks, and that the government shall establish sub-treasuries, or deposito-

[6] Dunning, *Op. Cit.*, pp. 163–165.

ries, in the several States; which sub-treasuries shall loan money to the people on approved security at a low rate of interest, not to exceed two per cent per annum: *Provided,* That real estate and non-perishable farm products shall be considered approved security; and that the circulating medium be increased to at least $50 per capita, keeping the volume equal to the demand.

For this the following substitute was adopted, to which Wade of Tennessee had his name withdrawn from this portion of the report:

1. *a.* We demand the abolition of national banks.

b. We demand that the government shall establish sub-treasuries or depositories in the several States, which shall loan money direct to the people at a low rate of interest, not to exceed two per cent per annum, on non-perishable farm products, and also upon real estate, with proper limitations upon the quantity of land and amount of money.

c. We demand that the amount of the circulating medium be speedily increased to not less than $50 per capita.

2. That we demand that Congress shall pass such laws as will effectually prevent the dealing in futures of all agricultural and mechanical productions; providing a stringent system of procedure in trials that will secure the prompt conviction, and imposing such penalties as shall secure the most perfect compliance with the law. Adopted.

3. We condemn the silver bill recently passed by Congress, and demand in lieu thereof the free and unlimited coinage of silver. Adopted.

4. We demand the passage of laws prohibiting alien ownership of land, and that Congress take prompt action to devise some plan to obtain all lands now owned by aliens and foreign syndicates; and that all lands now held by railroads and other corporations, in excess of such as is actually used and needed by them, be reclaimed by the government, and held for actual settlers only. Adopted.

5. Believing in the doctrine of equal rights to all, and special privileges to none, we demand:

a. That our national legislation shall be so framed in the future as not to build up one industry at the expense of another.

b. We further demand a removal of the existing heavy tariff tax from the necessities of life that the poor of our land must have.

c. We further demand a just and equitable system of graduated tax on incomes.

d. We believe that the money of the country should be kept as much as possible in the hands of the people, and hence we demand that all national and State revenues shall be limited to the necessary expenses of the government, economically and honestly administered. Adopted.

6. We demand the most rigid, honest, and just State and national governmental control and supervision of the means of public communication and transportation; and if this control and supervision does not remove the abuse now existing, we demand the government ownership of such means of communication and transportation. Adopted.

7. We demand that the Congress of the United States submit an amendment to the Constitution, providing for the election of United States Senators by direct vote of the people of each State. Adopted.

11   FROM   *James Baird Weaver*
The Threefold Contention of Industry

James Baird Weaver, a veteran of the Civil War and an Iowa lawyer, left the Republican party to be elected to Congress in 1878 as a Greenbacker. He ran for president in 1880 on the Greenback ticket, and he early identified himself with the Alliance Movement. This article describes some of the concerns of the man who a few months later would be nominated for the presidency on the Populist or People's party ticket. In the election of 1892 he received nearly a million popular votes and 22 votes in the electoral college. In 1896 he was influential in ob-

SOURCE. *Arena,* 5:427–435 (March 1892) .

taining the Populist nomination of Bryan for president and in bringing about the fusion of the Populists and Democrats. See the Dictionary of American Biography *for a short biography.*

There are three fundamental questions pressing for solution in America. Indeed, they to-day challenge the attention of the whole civilized world. They are distinct and yet cognate, segregated though inseparable, and seem destined to advance *pari passu,* and to conquer together. United they form the triple issue of organized labor, which for magnitude and importance has never been equalled since man became the subject of civil government. They are the wheat which has been winnowed from the chaff on the threshing-floor of the century.

The patient, long-suffering people are at last aroused, and there is hurrying to and fro. They seem to have received marching orders from some mysterious source, and are moving out against the strongholds of oppression on three distinct lines of attack, but within supporting distance of each other. It is evident that a general engagement is but a short march ahead.

One army corps proposes to give battle for our firesides; for a foothold and for standing-room upon the earth. It has inscribed upon its banner, "This planet is the common inheritance of all the people! All men have a natural right to a portion of the soil! Down with monopoly and speculation in land!"

The second is marching to deliver those who sit in darkness, —the needy who cry, the poor also, and him that hath no helper. They seek to open wide the door of opportunity, and to throw back the iron gates which shut out from the bounties of nature the miserably clad, wretchedly housed, shivering, haggard, care-worn victims of adversity and slaves of debt. Upon its guidon is the tracing of a whip of cords, upraised by the hand of Justice above the heads of the money changers. The legend underneath reads, "Money is the creature of human law! We will issue it for ourselves! Down with usury! Liberty for the captives!"

The third is leading an attack to get possession of the highways and lines of communication which have been wrenched from the people, and which connect cities, distant communities and States with their base of supplies. This corps has inscribed upon its flag the battle cry, "Restoration of the public highways! They belong to the people, and shall not be controlled by private speculators!"

When Barak, after he and his people had suffered twenty years of oppression, overthrew Jabin and the captain of his Host, Deborah declared that the battle was from heaven; that "the stars in their courses fought against Sisera." And may we not reverently believe that the struggle of the oppressed people of our day, to reinvest themselves of their lands, their money, and their highways, is from heaven also?

The Constitution provides that "The United States shall guarantee to every State in this Union a Republican form of government." This language implies a permanent contract—a joint pledge on the part of the Federal and State governments united, to maintain Democratic institutions throughout all the States; the general government pledging its great power that the people shall not be deprived of the form, and the States undertaking, as to all matters within their jurisdiction, to make their local institutions Republican in spirit, substance, and administration. In other words, we have here a solemn declaraton of purpose: a guaranty to all the people that government, both State and national, shall be held strictly to its original and lofty function, that of security to the citizen "certain inalienable rights," which he received at the generous hand of his Creator, and which no government has the right to impair or permit to be impaired or taken away. The pledge is that this obligation shall never be departed from, not even in form.

These "inalienable rights" are, first, such as grow out of the relation of man to his Creator, and second, those which spring from his relation to organized society or government. The land question comes under the first subdivision.

Can it be denied that all men have a natural right to a portion of the soil? Is not the use of the soil indispensable to life? If so, is not the right of all men to the soil as sacred as their right to life itself? These propositions are so manifestly true as to lie beyond the domain of controversy. To deny them is to call in question the right of man to inhabit the earth.

Tested by those axioms, the startling wickedness of our whole land system—which operates to deprive the weakest members, and even a vast majority of community, of the power to secure homes for themselves and families, rendering them fugitives and outcasts, and forcing them to pay tribute to others for the right to live; that murderous system which permits the rich and powerful to reach out and wrench from the unfortunate their restingplace upon the planet, and to acquire title to unlimited areas of the earth,—is at once revealed in all its hideous and monstrous outlines. It also discloses to us the unwelcome truth that our government, which was instituted to secure to man the unmolested enjoyment of his inalienable rights, has been transformed into an organized force for the destruction of those rights. Ordained to protect life, it proclaims death; undertaking to insure liberty to the citizen, it decrees bondage; and having encouraged its confiding subjects to start in pursuit of happiness, it presses to their famished lips the bitter cup of disappointment.

Society may, in some respects, be compared to a great forest. We can no more construct a secure and flourishing commonwealth amidst a community of tenants than you can grow a thrifty forest disconnected from the soil. Both men and trees receive their strength and growth from the earth. One tree cannot gather food for another. Each takes from the earth its own nourishment. When it ceases to do so it must perish. And the moment you sever man from the soil and deprive him of the power to return and till the earth in his own right, the love of home perishes within him. He comes as a freeman, and is transformed into a predial slave. And hence, concerning the absorbing question of land reform, we contend that the child who is born while we are penning these thoughts, comes into the world clothed with all the natural rights which Adam possessed when he was the sole inhabitant of the earth. Liberty to occupy the soil in his own right, to till it unmolested, as soon as he has the strength to do so, and to live upon the fruits of his toil without paying tribute to any other creature, are among the most sacred and essential of these rights. Any state of society which deprives him of these natural and inalienable safeguards, is an organized rebellion against the providence of God, a conspiracy against human life, and menace to the peace of community. When complete readjustment shall come, as come it must quickly, it will proceed in accordance with this fundamental truth. The stone which the builders rejected will then become the head of the corner.

The money and transportation problems relate to the second class of inalienable rights above mentioned. But in our day they are so directly related to those conferred by the Creator as to be practically inseparable from them. They are the instrumentalities through which the natural rights of man are rendered available in organized society. Such, it is clear, was the conclusion of the Fathers when they incorporated into the Constitution the following among other far-reaching and sweeping provisions:

"Congress shall have power to regulate commerce with foreign nations and among the several States, and with the Indian tribes."

Whatever may be the meaning of this provision, it is certain that the framers of the Constitution regarded the power to be exercised as too important to be confided to the discretion of individuals or left to the control of the States. It is taken away from both, and grouped with those matters which are of national concern—things which require the united wisdom of the country to solve, and the constant exercise of its combined power to sustain and enforce.

When this clause was incorporated into the Constitution, the Union was composed of only thirteen States, grouped together along the Atlantic seaboard; and at that time our internal commerce was but trifling. To-day forty-four fixed stars and four minor planets shine out from our galaxy. Interstate commerce has become annually so vast as to baffle computation. Then we had but three million souls. We now number more than sixty-three millions. We have crowded the nineteenth century full of marvellous achievements; but during the last quarter of that time there seems to have been a studied effort in certain powerful circles to discredit our Declaration of Independence, and to circumvent all that was accomplished for individual rights by our war for self-government and our later struggle for emancipation. We have been vigilant concerning everything except human rights and constitutional safeguards, and have suffered injuries to be inflicted upon the great body of the people which a century of the wisest legislation possible cannot fully efface.

We will first consider this provision of the Constitution negatively, and point out some things which Congress may not do under this grant of power.

First, Congress cannot disavow the obligation which this provision imposes, retrocede it to the States, or surrender it to the various traffic associations. It cannot grant to individuals or corporations such control over the instruments of commerce as will place the great body of the people at the mercy of those individuals or corporations. It cannot so regulate commerce among the States as to compel the farmers of the Northwest to ship their produce to Chicago and New York when they wish to transport it to St. Louis and New Orleans. The Congress could not prescribe such discriminations in freight rates as would compel Western merchants and jobbers to purchase their supplies in Chicago or Philadelphia when they desire to buy at Des Moines or Omaha. Congress may not prescribe rules for the control of commerce among the States which are designed to bankrupt the merchants and manufacturers of one locality and to enrich those of another. It could not scheme to stimulate the growth of trade in one city or manufacturing centre and to destroy it in another. Congress cannot rightfully grant to individuals and syndicates such control over the public highways and facilities for interstate traffic as will enable them to concentrate the entire cattle trade of the continent into a single city, or number of cities, dominated by a combination of harpies and commercial bandits. It could not conspire with individuals to grant to them such rates of transportation as would build up a gigantic oil monopoly, and enable them to crush out all competing producers and refiners. It could not enter into a conspiracy with the great anthracite coal companies to afford them ample facilities to transport their product, and refuse like favors to competing companies.

If Congress should openly attempt to commit such outrages as these, an indignant people would sweep them from place and power like a torrent. If persisted in despite public sentiment, it would be regarded as a declaration that government had been dissolved, and the people would fly to arms as the only refuge from the atrocity.

The Fathers evidently foresaw that evils of this character would arise if the power to regulate commerce were left to individuals or to the States, and hence took it away and vested it exclusively in Congress. Apprehending that at some time localities might still attempt to levy tribute upon others, and that Congress itself might not always be disposed to act with fairness, the framers of the Constitution were careful to expressly declare that "No preference shall be given by any regulation of commerce or revenue to the ports of one State over those of another."

We will now consider the powers and corresponding duties which this provision confers and enjoins upon Congress.

Commerce among the States consists in the interchange of merchandise or other movable property on an extended scale between the people of the different States. It finds its chief expression in the instruments used in the exchange and trans-shipment of the same. These are three in number.

1. Money.
2. Facilities for transportation.
3. Facilities for the transmission of intelligence.

It will be readily seen that these instrumentalities are the indispensable factors in modern civilization, and relate directly to the acquisition and distribution of wealth, and hence to the tranquillity of society and the maintenance of personal rights. Faithfully wielded by the general government, they constitute a triple-plated armor, capable, if held steadily toward the foe, of turning aside the heaviest projectiles of tyranny, and broad enough to shield at all times the whole body of the people. With this view of the subject before our minds, the wisdom of the provision which vests this power exclusively in Congress, and which excludes the insatiable passion of avarice from any share in its exercise, becomes apparent to all.

How has Congress discharged this important trust, and with what effect upon Democratic institutions? It will be readily seen that within the limits of this paper we can only treat the subject suggestively. But the mere interrogation foreshadows the startling outlines of our national dilemma, and the prodigious growth of corporate power at once rises like an impassable mountain barrier before the mind. The whole trinity of commercial instruments have been seized by corporations, wrenched from Federal control, and are being used to crush out the inalienable rights of the people. They are interlocked by mutual interests, and advance together in their work of plunder and subjugation. They constantly do all those things which Congress could not do without exciting insurrection. They make war upon organized labor, and annually lay tribute upon a subjugated people greater than was ever exacted by any conqueror or military chieftain since man has engaged in the brutalities of war. They corrupt our elections, contaminate our legislatures, and pollute our courts of justice. They have grown to be stronger than the government; and the army of Pinkertons, which is ever at their bidding, is greater by several thousand than the standing army of the United States. Instead of the government controlling the corporations, the latter dominate every department of State. We may no longer look to Congress, as at present dominated, for the regulation of these facilities. That body is bent on farming out its sovereign power to individuals and corporations, to be used for personal gain.

Our national banking system is the result of a compact between Congress and certain speculative syndicates, Congress agreeing to exercise the power to create the money, to bestow it as a gift, and to enforce its circulation; while the syndicates are to determine the quantity, and say when it shall be issued and retired. No currency whatever can be issued under this law unless it is first called for by associated usurers, and then they may retire it again at pleasure. If they decline to call for its issue, the affliction must be borne. If issued, and speculators desire to destroy it, the disastrous sacrifice must be endured. The power of the government to issue lies dormant until evoked by a private syndicate. Then the money flows into their hands, not to be expended in business or paid out for labor, but to be loaned at usury on private account. It cannot be reached by any other citizen of the republic except as it may be borrowed of those favorities, who arbitrarily dispense it solely for personal gain. To obtain it, the borrower must pay to these dispensers of sovereign favor from six to twenty times as much (according to locality) as was paid by the first recipient. It is a fine exhibition of Democratic government to see our Treasury Department create the currency, bestow it as a gift upon money lenders, and then stand by with cruel indifference and witness the misfortunes, the sharp competition, and the afflictions of life drive the rest of its devoted subjects to the feet of these purse-proud barons as suppliants and beggars for extortionate, second-hand favors. This system was borrowed from the mother country, where it was planned to foster established nobility, distinctions of caste, and imperial and dynastic pretensions; and those who planned it have always been satisfied with its operation. This, then, is our situation:

For a home upon the earth, the poor must sue at the feet of the land speculator.

For our currency, we are remanded to the mercies of a gigantic money trust.

For terms upon which we may use the highways, we must consult the kings of the rail and their private traffic associations.

For rapid transit of information, we bow obligingly to a telegraph monopoly dominated by a single mind.

Our money, our facilities for rapid interstate traffic, the telegraph,—the three subtle messengers of our intensified and advanced civilization,—all appropriated and dominated by private greed; wage labor superseded by the invention of machinery, and the cast-off laborer forbidden to return to the earth and cultivate it in his own right; population rapidly increasing; highways lined with tramps; cities over-crowded and congested; rural districts mortgaged to the utmost limit, and largely cultivated by tenants; crime extending its cancerous roots into the very vitals of society; colossal fortunes arising like Alpine ranges alongside of an ever widening and deepening abyss of poverty; usury respectable, and God's law contemned; corporations formed by thousands to crowd out individuals in the sharp competition for money, and the trust to drive weak corporations to the wall.

Such are some of the evils which have given rise to the discontent now so universal throughout the Union. From the investigations which this unrest has awakened has been evolved the "Threefold Contention of Industry," covering the great questions of Land, Money, and Transportation. Should it be the subject of criticism or matter of astonishment that our industrial people feel compelled to organize for mutual and peaceful defence? That they are actuated by the purest motives and the highest behests of judgment and conscience in making their demands, cannot for one moment be called in question. They are conscious, also, that their contention is based upon the impregnable rock of the Constitution and intrenched in the decisions of our Court of Last Resort. They do not seek to interfere with the rights of others, but to protect their own; to rebuild constitutional safeguards which have been thrown down; to restore to the people their lawful control over the essential instruments of commerce, and to give vitality to those portions of our Great Charter which were framed for the common good of all.

Let it be understood that organized labor demands at the bar of public opinion a respectful hearing. It will ask for nothing which it does not believe to be right, and with less than justice it will not be content. Conscious that it hath its quarrel just, in the struggle to obtain its demands it will employ and it invites the use of only such weapons as are proper in the highest type of manly intellectual combat.

12 FROM *The People's Platform, 1892*

Both the long preamble and the specific planks of the 1892 Populists platform represent positions quite familiar to and generally supported by most of the Alliance men. The platform reflects much that had been agreed to in the Ocala Demands of 1890. Hicks traces the development of both the party and the platform in The Populist Revolt, *chapters six to nine.*

Assembled upon the 116th anniversary of the Declaration of Independence, the People's Party of America in their first national convention, invoking upon their action the blessing of Almighty God, put forth in the name and on behalf of the people of this country, the following preamble and declaration of principles:

SOURCE. K. H. Porter and D. B. Johnson, *National Party Platforms* (University of Illinois, 1961) pp. 89–91.

PREAMBLE

The conditions which surround us best justify our co-operation; we meet in the midst of a nation brought to the verge of
moral, political, and material ruin. Corruption dominates the
ballot-box, the Legislatures, the Congress, and touches even the
ermine of the bench. The people are demoralized; most of the
States have been compelled to isolate the voters at the polling
places to prevent universal intimidation and bribery. The newspapers are largely subsidized or muzzled, public opinion si-
lenced, business prostrated, homes covered with mortgages, labor impoverished, and the land concentrating in the hands of
capitalists. The urban workmen are denied the right to organize
for self-protection; imported pauperized labor beats down their
wages, a hireling standing army, unrecognized by our laws, is established to shoot them down, and they are rapidly degenerating
into European conditions. The fruits of the toil of millions are
boldly stolen to build up colossal fortunes for a few, unprecedented in the history of mankind; and the possessors of these, in
turn despise the Republic and endanger liberty. From the same
prolific womb of governmental injustice we breed the two great
classes—tramps and millionaires.

The national power to create money is appropriated to enrich
bond-holders; a vast public debt payable in legal tender currency
has been funded into gold-bearing bonds, thereby adding millions to the burdens of the people.

Silver, which has been accepted as coin since the dawn of history, has been demonetized to add to the purchasing power of
gold by decreasing the value of all forms of property as well as
human labor, and the supply of currency is purposely abridged to
fatten usurers, bankrupt enterprise, and enslave industry. A vast
conspiracy against mankind has been organized on two conti-
nents, and it is rapidly taking possession of the world. If not met
and overthrown at once, it forebodes terrible social convulsions,
the destruction of civilization, or the establishment of an absolute
despotism.

We have witnessed for more than a quarter of a century the
struggles of the two great political parties for power and plunder,
while grievous wrongs have been inflicted upon the suffering

people. We charge that the controlling influence dominating both these parties have permitted the existing dreadful conditions to develop without serious effort to prevent or restrain them. Neither do they now promise us any substantial reform. They have agreed together to ignore, in the coming campaign, every issue but one. They propose to drown the outcries of a plundered people with the uproar of a sham battle over the tariff, so that capitalists, corporations, national banks, rings, trusts, watered stock, the demonetization of silver and the oppressions of the usurers may all be lost sight of. They propose to sacrifice our homes, lives, and children on the altar of mammon; to destroy the multitude in order to secure corruption funds from the millionaires.

Assembled on the anniversary of the birthday of the nation, and filled with the spirit of the grand general and chief who established our independence, we seek to restore the government of the Republic to the hands of "the plain people," with which class it originated. We assert our purposes to be identical with the purposes of the National Constitution, to form a more perfect union and establish justice, insure domestic tranquillity, provide for the common defense, promote the general welfare, and secure the blessings of liberty for ourselves and our posterity.

We declare that this Republic can only endure as a free government while built upon the love of the whole people for each other and for the nation; that it cannot be pinned together by bayonets; that the civil war is over and that every passion and resentment which grew out of it must die with it, and that we must be in fact, as we are in name, one united brotherhood of freemen.

Our country finds itself confronted by conditions for which there is no precedent in the history of the world; our annual agricultural productions amount to billions of dollars in value, which must, within a few weeks or months be exchanged for billions of dollars' worth of commodities consumed in their production; the existing currency supply is wholly inadequate to make this exchange; the results are falling prices, the formation of combines and rings, the impoverishment of the producing class. We pledge ourselves that, if given power, we will labor to correct these evils by wise and reasonable legislation, in accordance with the terms of our platform.

We believe that the power of government—in other words, of the people—should be expanded (as in the case of the postal service) as rapidly and as far as the good sense of an intelligent people and the teachings of experience shall justify, to the end that oppression, injustice and poverty, shall eventually cease in the land.

While our sympathies as a party of reform are naturally upon the side of every proposition which will tend to make men intelligent, virtuous and temperate, we nevertheless regard these questions, important as they are, as secondary to the great issues now pressing for solution, and upon which not only our individual prosperity, but the very existence of free institutions depend; and we ask all men to first help us to determine whether we are to have a republic to administer, before we differ as to the conditions upon which it is to be administered, believing that the forces of reform this day organized will never cease to move forward, until every wrong is remedied, and equal rights and equal privileges securely established for all the men and women of this country.

PLATFORM

We declare, therefore,

First—That the union of the labor forces of the United States this day consummated shall be permanent and perpetual; may its spirit enter into all hearts for the salvation of the Republic and the uplifting of mankind.

Second—Wealth belongs to him who creates it, and every dollar taken from industry without an equivalent is robbery. "If any will not work, neither shall he eat." The interests of rural and civic labor are the same; their enemies are identical.

Third—We believe that the time has come when the railroad corporations will either own the people or the people must own the railroads, and should the government enter upon the work of owning and managing all railroads, we should favor an amendment to the Constitution by which all persons engaged in the government service shall be placed under a civil service regulation of the most rigid character, so as to prevent the increase of the power of the national administration by the use of such additional government employees.

FINANCE—We demand a national currency, safe, sound, and flexible, issued by the general government only, a full legal tender for all debts, public and private, and that without the use of banking corporations, a just, equitable and efficient means of distribution direct to the people, at a tax not to exceed 2 per cent per annum, to be provided as set forth by the sub-treasury plan of the Farmers' Alliance, or a better system; also by payments in discharge of its obligations for public improvements.

1. We demand free and unlimited coinage of silver and gold at the present legal ratio of 16 to 1.

2. We demand that the amount of circulating medium be speedily increased to not less than $50 per capita.

3. We demand a graduated income tax.

4. We believe that the money of the country should be kept as much as possible in the hands of the people, and hence we demand that all State and national revenues shall be limited to the necessary expenses of the government, economically and honestly administered.

5. We demand that postal savings banks be established by the government for the safe deposit of the earnings of the people and to facilitate exchange.

TRANSPORTATION—Transportation being a means of exchange and a public necessity, the government should own and operate the railroads in the interest of the people. The telegraph and telephone, like the post office system, being a necessity for the transmission of news, should be owned and operated by the government in the interest of the people.

LAND—The land, including all the natural sources of wealth, is the heritage of the people, and should not be monopolized for speculative purposes, and alien ownership of land should be prohibited. All land now held by railroads and other corporations in excess of their actual needs, and all lands now owned by aliens, should be reclaimed by the government and held for actual settlers only.

13 FROM   *Frank Basil Tracy*

           Rise and Doom of the Populist Party

The articles by Frank Basil Tracy and Charles S. Gleed, both newspapermen, offer opposing views on the nature, accomplishments, and prospects of the Populists. Tracy sees the movement as socialistic; Gleed does not. Both presumably viewed the same evidence, although neither pays much attention to what was happening in the South.

For what does this party really and exactly stand? There is much ignorance on this subject, which must be clearly understood in order to determine its strength and forecast its future. The doctrinal basis of Populism is socialism. Without that basis, the castles of Populism could never have been reared so high and so strong. The beliefs of this party are not new. They have often been controverted by argument and by practical tests. But they are especially strong because of their basis, laid by a generation of paternal acts since the War. The three reforms to which this party is pledged relate to the same matters which constituted the real grievances of the farmers: land, transportation and money. The party demands Government ownership of all land not held by actual settlers, Government ownership of all transportation facilities, and Government issue of all money by its fiat alone. The platform declaration on the subject of land is vague, and is a remarkable modification of the communistic ideas first preached by the leaders. With this plank few would quarrel, although the proposed reclamation of lands granted to the railways would be absurd, as nearly all the valuable land has been sold. The Government ownership of railways and other means of transportation is another of their tenets which is undergoing modification. It is still, however, a favorite hobby with thousands and is clearly a scheme of pure socialism. They do not seem to realize that the

SOURCE: *Forum, 16*:240–250 (October 1893) .

placing of the seven hundred thousand men now engaged with American railways alone in the hands of any political party, would make that party's dislodgement from power almost impossible and would ultimately lead to a despotism. Nor do they propose a way to secure the ten billion dollars necessary to acquire these railways, except possibly by peculiar and characteristic financial schemes. Indeed, it is marvelous how these men, no matter how ignorant and unlearned, will furnish readily and confidently solutions for all problems of finance—the most intricate, delicate and least understood of all Government concerns.

The Chief underlying principle of all Populist financial schemes is fiat money. Free silver, a sub-treasury, *etc.,* are purely incidental. It is the cardinal faith of Populism, without which no man can be saved, that money can be created by the Government, in any desired quantity, out of any substance, with no basis but itself; and that such money will be good and legal tender, the Government stamp, only, being required. Free silver will bring some relief, but nothing permanent so long as "contraction of the currency" is possible. We must increase the volume of our currency; that is the desideratum. The Government, say the Populists, which by Protection rolls wealth into the manufacturer's lap, which constructs great harbors, buildings and defences, which gave us free land, pensions, bounties, railways, and created greenbacks, can do anything to increase our money supply. Nothing can give a clearer idea of the Populist view of money than this illustration given to me lately by one of the ablest Populists in the West: "The money-market is like the pork-market in which John Cudahy lost his millions. Eastern financiers and gold-bugs are attempting to corner the money-market, just as Mr. Cudahy attempted to corner the pork-market. Mr. Cudahy failed because the supply of pork was beyond his estimation. Wall Street is succeeding because the supply of money is limited. We insist that the Government should increase the circulating medium to $50 *per capita* and keep it there. As fast as the plutocrats gather in the money, the Government should issue more money until the money-corner is broken." Assuming that this absurd and ludicrous comparison is correct, one cannot help inquiring where the value of money would go after such a corner were broken. It is quite evident that it would go where Mr. Cudahy's pork went.

On this point let me quote from Professor Nicholas Paine Gilman's "Socialism and the American Spirit," in which he says:

" 'What do we care for abroad?' was the ingenuous remark of an American Congressman, when the long experience of Europe in regard to cheap money was brought forward in opposition to some 'wild cat' scheme of his own. His error was plain enough. Human nature and gold and silver dollars and paper substitutes for them are fundamentally the same in their workings in Europe and America. There is no Ohio scheme of political economy worthy of respect from intelligent men; there is no Nevada scheme of finance better adapted to American soil than the knowledge painfully acquired by the greatest commercial nation of the Old World."

It does seem surprising, with John Law only a century removed and the Argentine lesson staring us in the face, that Populist financial notions can exist. It might seem easy to banish all such nonsense by a half-hour's exposition of history and a fifteen-minute lesson from a political economy primer. But these people will accept no political economy as reliable and no history as unbiased. Their text-books are Bellamy, Donnelly *et id genus omne*. The Populist faith in the "Gover'ment" is supreme. The Government is all-powerful and it ought to be all-willing. When a Populist debtor is approached by a creditor, his reply is actually often in these words: "I can't pay the debt until the Government gives me relief." This intervention or saving grace of the Government is a personal influence to him, a thing of life. What shall minister to a mind diseased like the Populist's? Only constitutional remedies.

The constituent elements of this party give significant hints as to its character. The rank and file are composed of honest, intelligent men, mild in language and demeanor. During the Omaha convention the writer met frequently and conversed with an old friend, a delegate from an Iowa county. He was a "logical" Populist. One could read and analyze the entire movement in that man's record, which has always been socialistic. When I first knew him he was a Granger, then he became successively a Greenbacker, a Prohibitionist and a Populist. He is a man in more than comfortable circumstances, intelligent, honest and a

Christian. It was during his early struggle for subsistence that he became inoculated with the socialist virus, and it remains with him. On the day the convention assembled he exclaimed with fervor, "This campaign is opening just like the first Lincoln campaign." Although the adjectives, "honest," "sincere," and "earnest," may be applied to the followers, their antonyms fit the leaders. At least ninety per cent of these candidates and exhorters are destitute of personal or political integrity. They are political vagabonds, slanderers and demagogues. Their records in their former homes are unsavory. All of them keep in the sore spots of their minds the sad memories of conventions in which they were old party candidates to whom came overwhelming disaster.

At one of the first Alliance Congressional conventions in 1890, it was dramatically asserted that one of the candidates was a lawyer. He escaped violence and recovered a portion of his strength by explaining that although he had been a lawyer, he was then disbarred! A Populist convention in 1892 almost nominated for lieutenant-governor a man to whom in former days a jury had been only too kind. Scores of these stump electricians have had lifelong records as mountebanks. General Weaver is an honest man, personally, but he has boxed the compass in politics, always ready warmly to embrace any party or "ism" in the loving ecstasy of political hunger. It has been repeatedly charged that many of these Populist leaders are in the secret employ of the corporations, and the evidence, in some cases, seems conclusive. And so on with almost the entire list. Exceptions may be made of Senators Kyle and Allen, and Mrs. Lease, who knows really little of her themes, but is earnest and honest, and has wonderful hortatory power.

It would be unfair to close this article without recounting the excellent results of the organization of this party in the Western States. Like all third parties, it has done the good work of breaking up old political rings and corrupt administrations, making a cleansing of the old parties imperative. In Kansas it accomplished what would have been an impossibility under previous conditions by electing as a Congressman-at-large an ex-Confederate, thus burying all sectional strife. Barring such circumstances as the Kansas legislative barricades, the State elevator

law in Minnesota, and the "State-railway-to-the-gulf" scheme of
the Nebraska legislature, the public acts of this party have been very
creditable. This is especially true in Nebraska, where the party
passed the Australian ballot law and the maximum freight-rate
law, an act reducing the extortionate freight-rates. Gigantic
forces were at work to accomplish that bill's defeat, but the Populists were assisted by the united Omaha press and the anti-monopolists in the two greater parties. No more glittering baits were
ever held out to induce men to swerve from their duty. The law
is now suspended by a temporary injunction, issued at the demand of Eastern stockholders, who desire to postpone to a later
time this long-deferred act of justice. But the delay will have no
permanent effect, except to make good campaign material for
the Populists in this year's campaign for the judicial offices. Other meritorious and just laws were passed by this legislature, directed by Populists. In spite of all the great pressure of the corporations, the Populists, by the aid of the Democrats, elected to
the Senate an honest man, William Vincent Allen, against the
chosen friend of the monopolists, the Republican candidate, the
general solicitor of the Union Pacific railway. That election cost
Mr. Allen just $74.25. This was probably the smallest sum by
which a seat in the present United States Senate was secured.
Mr. Allen is a Populist, with a head filled with wrong financial
notions; but he is a conservative, pure, incorruptible man, who
won renown as an eminent attorney and a just, upright judge,
whose acts of kindness and charity are legion.

But the greatest benefit derived from this party's birth has
been educational. The whole country has been filled with the desire and spirit of investigation, and questions respecting finance
and Governmental functions have been studied by men and
women as they are studied nowhere else in the world. Out of this
Populist movement are gradually evolving sound arguments to
counteract their fallacies, and in this fact lies the very means for
accomplishing the party's overthrow.

It would be contrary to all reason, logic and history to presume an extended life for this party. As now constituted, it must
fail and fall, because it rests upon error. A significant index to its
decay is contained in the remarkable speed with which it is de-

serting its former tenets and growing in the direction of conservatism and sanity. Fiatism is mild compared with the wild plans of the enforced division of property, the spoliation of millionaires' estates, and the thousand other communistic and anarchistic schemes of the first campaign. The party seems intent on reaching safer ground. Although it would not consider a lawyer as a candidate for Congress in 1890, it elected a lawyer to the United States Senate in 1893. The sub-treasury and Government warehouse schemes have been virtually abandoned. A few weeks ago, Nebraska's most honored Populist in a public address defended the banks. This would have been a treasonable act three years ago. At the present time, the party is growing by the accretions of free-silver Democrats who feel that they have been betrayed by their party and their Executive. Could a Congressional election be held to-day, the Populists would win notable victories. But this growth will be temporary, just as the free-silver frenzy must soon pass away. Parenthetically, let it be understood that Populism will derive no benefit from the Waites, Wolcotts, Stewarts and other monometallists of the silver States. Their creed is purely selfish and is not sincere. This temporary agitation and the perfervid cries of "Western Empire" need alarm no Eastern mind, for their influence is very slight and does not sensibly affect the real Populists. Genuine Populism has its home in Minnesota, the Dakotas, Iowa, Nebraska, Missouri and Kansas. These States contain three times the population of the entire country West of them. What the Populists want is not free silver, *per se,* but more money. And as they have abandoned all their other tenets and staked their existence on fiat, foundationless currency, as a national party the People's Party must fail, just as the Greenback Party failed. Although as Mr. Reed says, humanity is incapable of prolonged virtue, yet the periods of financial heresy are always brief. And with the fiat money heresy dissipated, there will be no tie remaining to unite its elements into a national party.

Whether the party in the various Western States, in its growth toward sanity and reason, will retain strength enough to absorb a majority of the old parties, or whether its members will return to their former parties, are matters of mere conjecture, depending

upon local conditions. There are many influences at work to dis-
solve these State parties as well as the national party. One is the
educational impulse given by this movement, which is producing
arguments to refute its arguments. Another is the relatively comfortable condition of the Western farmers. Evidences of this dissolution are found in the decay of the Farmers' Alliances, nearly
half of them having given up their charters in Nebraska alone.
An analysis of last year's election figures will show that, as compared with 1890, the Populists made gains in the towns and lost
heavily in the country districts. This farmers' movement cannot
long endure when the farmers begin to desert it. Another influence is the recent financial depression, affecting farmers least
and "plutocrats" most. The conditions which would tend to pro-
long the life of the State parties are a continuance of real grievances, and hard times. The same local causes which immediately
preceded the uprising would keep it alive. Remedies must be
provided for railway and other extortions, and old party leaders
will be wise if they continue to remedy real grievances and thus
deprive the People's Party of a mission. There is always present
a great, overwhelming, unformed anti-monopoly spirit in the
West which, when excited to anger, will remedy monopolistic
wrongs within, if possible, and, if not, without party lines. But its
concern is with local and State affairs only.

So far as the national People's Party is concerned, since Populism is disappearing and its distinctive features are being lost,
its dissolution must soon come. The decay of the Western socialistic body, the parent of Populism, will not be so rapid. The deep
impressions of thirty years are not so easily removed. There is no
doubt, however, that many of the influences now destroying Populism will also affect the greater disease. The Populist movement has itself reacted against the spread of socialism by revealing to many who were regarding socialism favorably the logical
end of its doctrines in the foolish schemes of the new party. The
face of the nation, too, seems turned away from the enactment
and toward the repeal of paternal legislation. And as "looking to
Washington" still obtains, let Washington, by careful and wise
laws, teach the higher truths of individualism. The American
people should congratulate themselves that the discovery has so

soon been made and that a movement has begun to analyze and check this Western socialism which revealed its strength and, let us hope, attained its prime in the People's Party.

Frank Basil Tracy

14 FROM   *Charles S. Gleed*
The True Significance of Western Unrest

It is much believed in the East that recent political movements in the Western part of the United States are manifestations of socialism. I judge this belief to be not wholly, but chiefly, a consequence of the desire of those who entertain it. The socialists of the East, of every shade of belief, naturally desire the West to be with them. They have been quick to assert, therefore, that all the peculiar political activity in recent years in the West, with reference particularly to finance, wages, labor, taxation, and kindred subjects, has been socialistic. I am unable to see that this is true.

But some of those who believe that recent Western movements have been socialistic certainly do not desire to believe it. Such observers have probably been misled by reliance on the news columns of the daily press and hastily written editorial comments thereon. The news dispatches and reprints are fairly correct, so far as concerns the literal statement of fact. The impressions left, however, are often decidedly wrong. Thus the radical utterances of some street orator in Texas get reported in the local papers because a crowd of idle listeners are amused by the harangue and no better news demands space. By the time the local report is gravely quoted and commented on in the East, it presents a most serious aspect. Or, a mass meeting of five or ten men in Nebraska passes resolutions that would have gladdened the heart of Saint-Simon, and when the resolutions arrive in Boston they tell plainly of a full-fledged "movement." Or, again, the Populists of

SOURCE. *Forum, 16:*251–260 (October 1893) .

Kansas and Colorado, with their peculiar financial sleep-walking and their necromancy of grip, countersign and dark closet, by the magnifying power of intervening mists take on an appearance in New England like that of the most radical social reformers of Europe. Now add to these deceptive indices the bold claims made by Eastern socialistic reformers as to the progress made in the West by their several favorite theories, and it becomes natural that many should be misled.

It is true, of course, that there are in the West, as elsewhere, social reformers of all grades. The West has a cosmopolitan population. Nearly all her people "came from somewhere" and nearly all were started out of old ruts by some form of necessity. As it is true that necessity is not so very much of a stimulus to dull people, so it happens that the people of the West are mostly of a keen, active and sturdy mental type. They may be crude, but they are very much alive. These facts explain clearly a number of things. They explain why there is no belief entertained anywhere on the face of the earth which does not have its disciples in the West. From the Chinese of California, the Mormons of Utah, the Mennonites of Kansas, and the Schweinfurthians and Anarchists of Illinois, to the most erratic private doctrinaires, the West includes representatives of all that the intellectual world contains. But this is not admitting that any of the well-defined political movements of the West have socialism as a motive power.

Again, the character of Western people, as I have described it, explains why they are apt to "speak out in meeting" more often and more vigorously than their Eastern brethren. Western people are not only keenly alive, but their lives are so ordered that they have time and opportunity to read, think, debate, convene, and resolve. This they do with unfailing regularity and a high average of intelligence. Eastern people are divided in a way that Western people are not. Either they are prosperous and have few political griefs which they need to get excited about, or else they are handicapped in so many ways as to be incapable of much political action except of the mob kind. Not many speeches, articles, legislative bills, or organized reforms can be expected of the miners of Pennsylvania, the mill operatives of New England, the poorer class of tenant-farmers and coast-wise laborers of the At-

lantic coast, or the cheap laborers of the great Eastern cities. But
Western farmers and workers generally are trained so universally
not only to hear but to be heard, not only to read but to write,
that they arrive easily at that point of fluency where in times of
change and excitement they say and applaud that which they do
not fully believe. They have rarely made themselves widely ridic-
ulous by ill-considered utterances, and still more rarely by ill-
considered action; but when circumstances conspire greatly to ir-
ritate, there is no denying that a great crop of unripe opinions
and immature expressions is the result.

The recent political manifestations in the West may properly
be said to cover agitations relating to greenbacks, the coinage of
the precious metals, and other currency questions; proposed
changes in systems of state and national banking; defaults, repu-
diations and readjustments of debts; strikes, boycotts, unions,
leagues, and other demonstrations of organized labor; granger-
ism, labor-unionism, populism and other political or semi-politi-
cal organizations based on secret, fraternal and coöperative
methods; and finally, speculations, failures, alarms, legislations,
and migrations affecting the finances of the country. It is somewhat difficult to determine whether to be glad or sorry that these
movements cannot be charged to socialism. If they could be considered socialistic, there would be ground at least for respect for
the high moral purposes involved. As matters stand, there can be
discovered only the methods and motives of the ordinary political
reformer—whether leader or voter—the motives of desire to accomplish proper but comparatively small and superficial purposes. The motive of such a voter is to attain an immediate gain, or
cure an immediate annoyance. The motive of such a leader, the
office-seeker, is partly the same as that of the voter, and partly
that he may keep in line with the voter.

The old Granger movement was really more practical in a socialistic direction than the People's party movement. The Grange
was founded on the theory of the love of all for all, and it under-
took to do paternal work in a commercial direction by establishing union stores, abolishing intermediate profits between producers and consumers, *etc.*

The demand for the free coinage of silver (no matter whether
it is a just demand or not), as it is now made under the leader-

ship of the silver-producing States, most certainly shows no signs
of a socialistic origin. This demand is made by those who own
silver mines; those who labor in silver mines; those whose business is directly with the owners of and laborers in silver mines;
those who believe that free coinage means free coin—at least to
an extent which will enable them to get some; those who believe
that there has actually been a contraction of the currency hurtful
to the debtor class, which contraction free coinage will cure;
those who sincerely believe in the possibility of a double standard; and, finally, those seeking to get or retain office, who believe
the free silver side the safest one for them to take. All these
classes are practically content with things in general as they now
are. They belong to the Republican, the Democratic, or the other
parties, as a rule, and are vigorous partisans. Senator Wolcott, a
Republican, Senator Peffer, a Populist, and Senator Martin, a
Democrat, go arm-in-arm near the head of the free silver proces-
sion. All three would quickly resent any charge that they are socialists.

In all the proposed changes of banking systems, nothing has
appeared more radical than the semi-banking scheme of the Populists of two years ago in their demand for sub-treasuries. This
demand had a socialistic look, inasmuch as it meant that the government should become practically the purchaser, or at least the
guarantor, of the agricultural products of the country. But the cry
for the sub-treasury cannot now be heard even in the silence of
the night. When the People's party of the last Kansas legislature
was listening to competitive speeches from candidates for the
United States Senate, the men who claimed to be in favor of
sub-treasuries were turned down for a life-long Democrat who
told the caucus plainly that he was not for the sub-treasury
scheme and did not believe his hearers were.

The Populists, in their various professions, have, up to date,
shown signs of a strong belief in a paternal government, with all
which that implies. They have wanted the Government to do everything, from storing grain, lending money, and owning railroads, to fixing hours of labor and the value of the commodities
of commerce. This is why the public first laughed, then wept, at
the grotesque spectacle of the Democratic party, in some of the
Western States, surrendering its individuality to the Populist par-

ty in a fusion which became a merger. But the real paternalists
of the People's party, if there are any, are the leaders who draw
the platforms. The voters, so far in the party's history, have not
known what they were doing. It is not socialism they have been
pursuing. Rather it is hard times that they have been trying to escape. The demagogues who wanted offices have not dared to
preach old-fashioned truth. They have not dared to advocate
economy, a reduction of debts by the payment of them, a diversi-
fication of industries, and such time-honored remedies for a sick
body politic; so they have advocated sundry specious doctrines
nearly enough practicable to promise the people everything and
far enough from practicable to be free of all danger of ever being
tried. The members of the People's party, considered as a whole,
would deny firmly that they are socialists. They say if the Republican party were only as good as it once was, they would vote
with it. They do not ask a new Heaven or a new earth. All they
want is more money. They are fast getting it, and the faster they
get it the more reluctant they become to ride forty miles in a
lumber-wagon through the rain to hear Mrs. Lease and General
Weaver make speeches.

All phases of this discussion lead up to the subject of debt. In
all stages of the development of the West there have been great
mutations in the status of public and private debt. There have all
along been periods of general default, succeeding times of readjustment, with here and there a clear case of repudiation. The
commercial world seems to have no difficulty in distinguishing
clearly between readjustment and repudiation. Every reorganiza-
tion committee in New York and London bears witness to the
propriety of readjustment under certain circumstances. Necessary readjustments have sometimes been wrongly called repudiation, though probably not often. Defaults and readjustments have
recently characterized the business and financial history of the
West. Much of all that is now immediately before us politically is
due to this fact. For some years, all the Western country, including its southern States, was the field of great development and far
greater speculation. Confidence in the West was unlimited, cau-
tion about it was uncommon, recklessness concerning it was
prevalent. When the tide turned, Eastern capitalists, with equally
reckless haste, withdrew all money possible from the West. This,

added to already existing hardships, made the situation unprecedented and led to some real and much threatened improper legislation. This made a bad situation only worse, and the West was soon deprived of the capital which it had almost a moral right to depend on in furtherance of its legitimate growth. The Eastern capitalists having, as they felt, escaped from the West, turned with their money in hand to their near neighbors and bought at random whatever glittering bubbles the latter had to sell. In due time, and perhaps a little sooner, the bubbles did as bubbles do —they burst. "The last condition of this man was worse than the first." After pain in the West and agony in the East, what else could he do than hide his remnant in a strong-box? Then arose a general alarm among manufacturers for fear the party in power would keep its promise to slaughter the tariff. Merchants withheld their orders and the mills shut down. In addition to this, it may be safely asserted that the usual somewhat indefinite, but inevitable period, had arrived for the cancellation of part of the great volume of outstanding promises to pay in the commercial world at large. In Great Britain and all the fields of her vast financial ventures, including our own country, the time had come —and it has not yet altogether passed—for the general cancellation of hopeless promises to pay. All this and more brought distrust to all the people, and their dollars hid like frightened mice. Solvent banks broke and worthy institutions surrendered. Verdict: "A lack of confidence."

To this provocation of financial pressure are due the very recent financial manifestations in the West. But in them socialism has really not appeared.

It is true that the farmers, especially the farmers of the Farmers' Alliance and the People's Party, talk much in these times of union and coöperation; but in practice they all compete most diligently, and each is clearly and undeniably anxious for the private ownership of land and everything else—each man to be, as far as possible, himself the private owner. Every agency ever suggested by the People's party for the improvement of the condition of its members has been designed solely for the assistance of limited classes and of their attempts to prosper by present methods, without reference to the condition of those in other occupations.

In brief, every demand for a change in the financial methods of the nation which has seemed to come in late years from the West, has been based almost wholly on the general financial embarrassment of many people. These people, so far as the West is concerned, have not been and are not prodigals, tramps or loafers. On the contrary, they have been and are people of personal worth. They have striven steadily for years to accumulate money and have failed. They are no doubt over-receptive to explanations of the causes of their failure, and incline too readily to any plausible remedy that is offered. But it still remains true that they are as satisfied as people ever were with the world as it is, and would be and continue practically happy if they could get fair wages and reasonable profits, and get them steadily. They are not conscious of having been to any material extent the authors of their own misfortunes, as is many times the case. All they ask is the society they have always known, with the prosperity they have from time to time enjoyed. This is not socialism.

Charles S. Gleed

15 FROM K. H. Porter and D. B. Johnson Party Platforms of 1896

The ancient witticism that party platforms are something to run on, not to stand on, seems to have a kind of immortality, perhaps, because it does harbor a grain or more of truth. But at the time of their adoption, party platforms, aside from the elaborate denunciation of the opposition, often represent a core of beliefs or commitments to which the majority of the delegates subscribe. Thus the 1896 platforms of the Republicans, the Democrats, and the Populists are worthy of careful examination, and both the Democratic and the Populist platforms should be compared with the Populist platform of 1892 and the grievances and demands of the Alliances of the 1880s.

SOURCE. K. H. Porter and D. B. Johnson, *National Party Platforms*, University of Illinois, 1961, pp. 97–109.

DEMOCRATIC PLATFORM OF 1896

We, the Democrats of the United States in National Convention assembled, do reaffirm our allegiance to those great essential principles of justice and liberty, upon which our institutions are founded, and which the Democratic Party has advocated from Jefferson's time to our own—freedom of speech, freedom of the press, freedom of conscience, the preservation of personal rights, the equality of all citizens before the law, and the faithful observance of constitutional limitations.

During all these years the Democratic Party has resisted the tendency of selfish interests to the centralization of governmental power, and steadfastly maintained the integrity of the dual scheme of government established by the founders of this Republic of republics. Under its guidance and teachings the great principle of local self-government has found its best expression in the maintenance of the rights of the States and in its assertion of the necessity of confining the general government to the exercise of the powers granted by the Constitution of the United States.

The Constitution of the United States guarantees to every citizen the rights of civil and religious liberty. The Democratic Party has always been the exponent of political liberty and religious freedom, and it renews its obligations and reaffirms its devotion to these fundamental principles of the Constitution.

The Money Plank

Recognizing that the money question is paramount to all others at this time, we invite attention to the fact that the Federal Constitution named silver and gold together as the money metals of the United States, and that the first coinage law passed by Congress under the Constitution made the silver dollar the monetary unit and admitted gold to free coinage at a ratio based upon the silver-dollar unit.

We declare that the act of 1873 demonetizing silver without the knowledge or approval of the American people has resulted in the appreciation of gold and a corresponding fall in the prices of commodities produced by the people; a heavy increase in the burdens of taxation and of all debts, public and private; the enrichment of the money-lending class at home and abroad; the prostration of industry and impoverishment of the people.

We are unalterably opposed to monometallism which has locked fast the prosperity of an industrial people in the paralysis of hard times. Gold monometallism is a British policy, and its adoption has brought other nations into financial servitude to London. It is not only un-American but anti-American, and it can be fastened on the United States only by the stifling of that spirit and love of liberty which proclaimed our political independence in 1776 and won it in the War of the Revolution.

We demand the free and unlimited coinage of both silver and gold at the present legal ratio of 16 to 1 without waiting for the aid or consent of any other nation. We demand that the standard silver dollar shall be a full legal tender, equally with gold, for all debts, public and private, and we favor such legislation as will prevent for the future the demonetization of any kind of legal-tender money by private contract.

We are opposed to the policy and practice of surrendering to the holders of the obligations of the United States the option reserved by law to the Government of redeeming such obligations in either silver coin or gold coin.

Interest-Bearing Bonds

We are opposed to the issuing of interest-bearing bonds of the United States in time of peace and condemn the trafficking with banking syndicates, which, in exchange for bonds and at an enormous profit to themselves, supply the Federal Treasury with gold to maintain the policy of gold monometallism.

Against National Banks

Congress alone has the power to coin and issue money, and President Jackson declared that this power could not be delegated to corporations or individuals. We therefore denounce the issuance of notes intended to circulate as money by National banks as in derogation of the Constitution, and we demand that all paper which is made a legal tender for public and private debts, or which is receivable for dues to the United States, shall be issued by the Government of the United States and shall be redeemable in coin.

Tariff Resolution

We hold that tariff duties should be levied for purposes of revenue, such duties to be so adjusted as to operate equally throughout the country, and not discriminate between class or section, and that taxation should be limited by the needs of the Government, honestly and economically administered. We denounce as disturbing to business the Republican threat to restore the McKinley law, which has twice been condemned by the people in National elections and which, enacted under the false plea of protection to home industry, proved a prolific breeder of trusts and monopolies, enriched the few at the expense of the many, restricted trade and deprived the producers of the great American staples of access to their natural markets.

Until the money question is settled we are opposed to any agitation for further changes in our tariff laws, except such as are necessary to meet the deficit in revenue caused by the adverse decision of the Supreme Court on the income tax. But for this decision by the Supreme Court, there would be no deficit in the revenue under the law passed by the Democratic Congress in strict pursuance of the uniform decisions of that court for nearly 100 years, that court having in that decision sustained Constitutional objections to its enactment which had previously been over-ruled by the ablest Judges who have ever sat on that bench. We declare that it is the duty of Congress to use all the Constitutional power which remains after that decision, or which may come from its reversal by the court as it may hereafter be constituted, so that the burdens of taxation may be equally and impartially laid, to the end that wealth may bear its due proportion of the expense of the Government.

Immigration and Arbitration

We hold that the most efficient way of protecting American labor is to prevent the importation of foreign pauper labor to compete with it in the home market, and that the value of the home market to our American farmers and artisans is greatly reduced by a vicious monetary system which depresses the prices of their products below the cost of production, and thus deprives them of the means of purchasing the products of our home manufactories; and as labor creates the wealth of the country, we demand the passage of such laws as may be necessary to protect it in all its rights.

We are in favor of the arbitration of differences between employers engaged in interstate commerce and their employees, and recommend such legislation as is necessary to carry out this principle.

Trusts and Pools

The absorption of wealth by the few, the consolidation of our leading railroad systems, and the formation of trusts and pools require a stricter control by the Federal Government of those arteries of commerce. We demand the enlargement of the powers of the Interstate Commerce Commission and such restriction and guarantees in the control of railroads as will protect the people from robbery and oppression.

Declare for Economy

We denounce the profligate waste of the money wrung from the people by oppressive taxation and the lavish appropriations of recent Republican Congresses, which have kept taxes high, while the labor that pays them is unemployed and the products of the people's toil are depressed in price till they no longer repay the cost of production. We demand a return to that simplicity and economy which befits a Democratic Government, and a reduction in the number of useless offices, the salaries of which drain the substance of the people.

Federal Interference in Local Affairs

We denounce arbitrary interference by Federal authorities in local affairs as a violation of the Constitution of the United

States, and a crime against free institutions, and we especially object to government by injunction as a new and highly dangerous form of oppression by which Federal Judges, in contempt of the laws of the States and rights of citizens, become at once legislators, judges and executioners; and we approve the bill passed at the last session of the United States Senate, and now pending in the House of Representatives, relative to contempts in Federal courts and providing for trials by jury in certain cases of contempt.

Pacific Railroad

No discrimination should be indulged in by the Government of the United States in favor of any of its debtors. We approve of the refusal of the Fifty-third Congress to pass the Pacific Railroad Funding bill and denounce the effort of the present Republican Congress to enact a similar measure.

Pensions

Recognizing the just claims of deserving Union solidiers, we heartily indorse the rule of the present Commissioner of Pensions, that no names shall be arbitrarily dropped from the pension roll; and the fact of enlistment and service should be deemed conclusive evidence against disease and disability before enlistment.

Admission of Territories

We favor the admission of the Territories of New Mexico, Arizona and Oklahoma into the Union as States, and we favor the early admission of all the Territories, having the necessary population and resources to entitle them to Statehood, and, while they remain Territories, we hold that the officials appointed to administer the government of any Territory, together with the District of Columbia and Alaska, should be *bona-fide* residents of the Territory or District in which their duties are to be performed. The Democratic party believes in home rule and that all public lands of the United States should be appropriated to the establishment of free homes for American citizens.

We recommend that the Territory of Alaska be granted a delegate in Congress and that the general land and timber laws of the United States be extended to said Territory.

Sympathy for Cuba

The Monroe doctrine, as originally declared, and as interpreted by succeeding Presidents, is a permanent part of the foreign policy of the United States, and must at all times be maintained.

We extend our sympathy to the people of Cuba in their heroic struggle for liberty and independence.

Civil-Service Laws

We are opposed to life tenure in the public service, except as provided in the Constitution. We favor appointments based on merit, fixed terms of office, and such an administration of the civil-service laws as will afford equal opportunities to all citizens of ascertained fitness.

Third-Term Resolution

We declare it to be the unwritten law of this Republic, established by custom and usage of 100 years, and sanctioned by the examples of the greatest and wisest of those who founded and have maintained our Government that no man should be eligible for a third term of the Presidential office.

Improvement of Waterways

The Federal Government should care for and improve the Mississippi River and other great waterways of the Republic, so as to secure for the interior States easy and cheap transportation to tidewater. When any waterway of the Republic is of sufficient importance to demand aid of the Government such aid should be extended upon a definite plan of continuous work until permanent improvement is secured.

Conclusion

Confiding in the justice of our cause and the necessity of its success at the polls, we submit the foregoing declaration of principles and purposes to the considerate judgment of the American people. We invite the support of all citizens who approve them and who desire to have them made effective through legislation, for the relief of the people and the restoration of the country's prosperity.

PEOPLE'S PLATFORM OF 1896

The People's Party, assembled in National Convention, reaffirms its allegiance to the principles declared by the founders of the Republic, and also to the fundamental principles of just government as enunciated in the platform of the party in 1892.

We recognize that through the connivance of the present and preceding Administrations, the country has reached a crisis in its National Life, as predicted in our declaration four years ago, and that prompt and patriotic action is the supreme duty of the hour.

We realize that, while we have political independence, our financial and industrial independence is yet to be attained by restoring to our country the Constitutional control and exercise of the functions necessary to a people's government, which functions have been basely surrendered by our public servants to corporate monopolies. The influence of European moneychangers has been more potent in shaping legislation than the voice of the American people. Executive power and patronage have been used to corrupt our legislatures and defeat the will of the people, and plutocracy has thereby been enthroned upon the ruins of democracy. To restore the Government intended by the fathers, and for the welfare and prosperity of this and future generations, we demand the establishment of an economic and financial system which shall make us masters of our own affairs and independent of European control, by the adoption of the following declaration of principles:

The Finances

1. We demand a National money, safe and sound, issued by the General Government only, without the intervention of banks of issue, to be a full legal tender for all debts, public and private; a just, equitable, and efficient means of distribution, direct to the people, and through the lawful disbursements of the Government.

2. We demand the free and unrestricted coinage of silver and gold at the present legal ratio of 16 to 1, without waiting for the consent of foreign nations.

3. We demand that the volume of circulating medium be speedily increased to an amount sufficient to meet the demand of the business and population, and to restore the just level of prices of labor and production.

4. We denounce the sale of bonds and the increase of the public interest-bearing debt made by the present Administration as unnecessary and without authority of law, and demand that no more bonds be issued, except by specific act of Congress.

5. We demand such legislation as will prevent the demonetization of the lawful money of the United States by private contract.

6. We demand that the Government, in payment of its obligation, shall use its option as to the kind of lawful money in which they are to be paid, and we denounce the present and preceding Administrations for surrendering this option to the holders of Government obligations.

7. We demand a graduated income tax, to the end that aggregated wealth shall bear its just proportion of taxation, and we regard the recent decision of the Supreme Court relative to the income-tax law as a misinterpretation of the Constitution and an invasion of the rightful powers of Congress over the subject of taxation.

8. We demand that postal savings-banks be established by the Government for the safe deposit of the savings of the people and to facilitate exchange.

Railroads and Telegraphs

1. Transportation being a means of exchange and a public necessity, the Government should own and operate the railroads in the interest of the people and on a non-partisan basis, to the end that all may be accorded the same treatment in transportation, and that the tyranny and political power now exercised by the great railroad corporations, which result in the impairment, if not the destruction of the political rights and personal liberties of the citizens, may be destroyed. Such ownership is to be accomplished gradually, in a manner consistent with sound public policy.

2. The interest of the United States in the public highways built with public moneys, and the proceeds of grants of land to the Pacific railroads, should never be alienated, mortgaged, or sold, but guarded and protected for the general welfare, as provided by the laws organizing such railroads. The foreclosure of existing liens of the United States on these roads should at once follow default in the payment thereof by the debtor companies; and at the foreclosure sales of said roads the Government shall purchase the same, if it becomes necessary to protect its interests therein, or if they can be purchased at a reasonable price; and the Government shall operate said railroads as public highways for the benefit of the whole people, and not in the interest of the few, under suitable provisions for protection of life and property, giving to all transportation interests equal privileges and equal rates for fares and freight.

3. We denounce the present infamous schemes for refunding these debts, and demand that the laws now applicable thereto be executed and administered according to their intent and spirit.

4. The telegraph, like the Post Office system, being a necessity for the transmission of news, should be owned and operated by the Government in the interest of the people.

The Public Lands

1. True policy demands that the National and State legislation shall be such as will ultimately enable every prudent and industrious citizen to secure a home, and therefore the land should not

be monopolized for speculative purposes. All lands now held by railroads and other corporations in excess of their actual needs should by lawful means be reclaimed by the Government and held for actual settlers only, and private land monopoly, as well as alien ownership, should be prohibited.

2. We condemn the land grant frauds by which the Pacific railroad companies have, through the connivance of the Interior Department, robbed multitudes of *bona-fide* settlers of their homes and miners of their claims, and we demand legislation by Congress which will enforce the exemption of mineral land from such grants after as well as before the patent.

3. We demand that *bona-fide* settlers on all public lands be granted free homes, as provided in the National Homestead Law, and that no exception be made in the case of Indian reservations when opened for settlement, and that all lands not now patented come under this demand.

The Referendum

We favor a system of direct legislation through the initiative and referendum, under proper Constitutional safeguards.

Direct Election of President and Senators By the People

We demand the election of President, Vice-President, and United States Senators by a direct vote of the people.

Sympathy for Cuba

We tender to the patriotic people of Cuba our deepest sympathy for their heroic struggle for political freedom and independence, and we believe the time has come when the United States, the great Republic of the world, should recognize that Cuba is, and of right ought to be, a free and independent state.

The Territories

We favor home rule in the Territories and the District of Columbia, and the early admission of the Territories as States.

Public Salaries

All public salaries should be made to correspond to the price of labor and its products.

Employment to be Furnished by Government

In times of great industrial depression, idle labor should be employed on public works as far as practicable.

Arbitrary Judicial Action

The arbitrary course of the courts in assuming to imprison citizens for indirect contempt and ruling by injunction should be prevented by proper legislation.

Pensions

We favor just pensions for our disabled Union soldiers.

A Fair Ballot

Believing that the elective franchise and an untrammeled ballot are essential to a government of, for, and by the people, the People's party condemns the wholesale system of disfranchisement adopted in some States as unrepublican and undemocratic, and we declare it to be the duty of the several State legislatures to take such action as will secure a full, free and fair ballot and an honest count.

The Financial Question "The Pressing Issue"

While the foregoing propositions constitute the platform upon which our party stands, and for the vindication of which its organization will be maintained, we recognize that the great and pressing issue of the pending campaign, upon which the present election will turn, is the financial question, and upon this great and specific issue between the parties we cordially invite the aid and co-operation of all organizations and citizens agreeing with us upon this vital question.

REPUBLICAN PLATFORM OF 1896

The Republicans of the United States, assembled by their representatives in National Convention, appealing for the popular and historical justification of their claims to the matchless achievements of thirty years of Republican rule, earnestly and confidently address themselves to the awakened intelligence, experience and conscience of their countrymen in the following declaration of facts and principles:

For the first time since the civil war the American people have witnessed the calamitous consequence of full and unrestricted Democratic control of the government. It has been a record of unparalleled incapacity, dishonor and disaster. In administrative management it has ruthlessly sacrificed indispensable revenue, entailed an unceasing deficit, eked out ordinary current expenses with borrowed money, piled up the public debt by $262,000,000 in time of peace, forced an adverse balance of trade, kept a perpetual menace hanging over the redemption fund, pawned American credit to alien syndicates and reversed all the measures and results of successful Republican rule. In the broad effect of its policy it has precipitated panic, blighted industry and trade with prolonged depression, closed factories, reduced work and wages, halted enterprise and crippled American production, while stimulating foreign production for the American market. Every consideration of public safety and individual interest demands that the government shall be wrested from the hands of those who have shown themselves incapable of conducting it without disaster at home and dishonor abroad and shall be restored to the party which for thirty years administered it with unequaled success and prosperity. And in this connection, we heartily endorse the wisdom, patriotism and success of the administration of Benjamin Harrison. We renew and emphasize our allegiance to the policy of protection, as the bulwark of American industrial independence, and the foundation of American development and prosperity. This true American policy taxes foreign products and encourages home industry. It puts the burden of revenue on foreign goods; it secures the American market for the American producer. It upholds the American standard of wages for the American workingman; it puts the factory by the side of the farm and makes the American farmer less dependent on foreign demand and price; it diffuses general thrift, and founds the strength of all on the strength of each. In its reasonable application it is just, fair and impartial, equally opposed to foreign control and domestic monopoly to sectional discrimination and individual favoritism.

We denounce the present tariff as sectional, injurious to the public credit and destructive to business enterprise. We demand such an equitable tariff on foreign imports which come into competition with the American products as will not only furnish adequate revenue for the necessary expenses of the Government, but will protect American labor from degradation and the wage level of other lands. We are not pledged to any particular schedules. The question of rates is a practical question, to be governed by the conditions of time and of production. The ruling and uncompromising principle is the protection and development of American labor and industries. The country demands a right settlement, and then it wants rest.

We believe the repeal of the reciprocity arrangements negotiated by the last Republican Administration was a National calamity, and demand their renewal and extension on such terms as will equalize our trade with other nations, remove the restrictions which now obstruct the sale of American products in the ports of other countries, and secure enlarged markets for the products of our farms, forests, and factories.

Protection and Reciprocity are twin measures of American policy and go hand in hand. Democratic rule has recklessly struck down both, and both must be re-established. Protection for what we produce; free admission for the necessaries of life which we do not produce; reciprocal agreement of mutual interests, which gain open markets for us in return for our open markets for others. Protection builds up domestic industry and trade and secures our own market for ourselves; reciprocity builds up foreign trade and finds an outlet for our surplus. We condemn the present administration for not keeping pace [faith] with the sugar producers of this country. The Republican party favors such protection as will lead to the production on American soil of all the sugar which the American people use, and for which they pay other countries more than one hundred million dollars annually. To all our products; to those of the mine and the fields, as well as to those of the shop and the factory, to hemp and wool, the product of the great industry sheep husbandry; as well as to the foundry, as to the mills, we promise the most ample protection. We favor the early American policy of discriminating duties for the upbuilding of our merchant marine. To the protection of our shipping in the foreign-carrying trade, so that American ships, the product of American labor, employed in American ship-yards, sailing under the stars and stripes, and manned, officered and owned by Americans, may regain the carrying of our foreign commerce.

The Republican party is unreservedly for sound money. It caused the enactment of a law providing for the redemption [resumption] of specie payments in 1879. Since then every dollar has been as good as gold. We are unalterably opposed to every measure calculated to debase our currency or impair the credit of our country. We are therefore opposed to the free coinage of silver, except by international agreement with the leading commercial nations of the earth, which agreement we pledge ourselves to promote, and until such agreement can be obtained the existing gold standard must be maintained. All of our silver and paper currency must be maintained at parity with gold, and we favor all measures designated to maintain inviolable the obligations of the United States, of all our money, whether coin or paper, at the present standard, the standard of most enlightened nations of the earth.

The veterans of the Union Armies deserve and should receive fair treatment and generous recognition. Whenever practicable they should be given the preference in the matter of employment. And they are entitled to the enactment of such laws as are best calculated to secure the fulfillment of the pledges made to them in the dark days of the country's peril.

We denounce the practice in the pension bureau so recklessly and unjustly carried on by the present Administration of reducing pensions and arbitrarily dropping names from the rolls, as deserving the severest condemnation of the American people.

Our foreign policy should be at all times firm, vigorous and dignified, and all our interests in the western hemisphere should be carefully watched and guarded.

The Hawaiian Islands should be controlled by the United States, and no foreign power should be permitted to interfere with them. The Nicaragua Canal should be built, owned and operated by the United States. And, by the purchase of the Danish Islands we should secure a much needed Naval station in the West Indies.

The massacres in Armenia have aroused the deep sympathy and just indignation of the American people, and we believe that the United States should exercise all the influence it can properly exert to bring these atrocities to an end. In Turkey, American residents have been exposed to gravest [grievous] dangers and American property destroyed. There, and everywhere, American citizens and American property must be absolutely protected at all hazards and at any cost.

We reassert the Monroe Doctrine in its full extent, and we reaffirm the rights of the United States to give the Doctrine effect by responding to the appeal of any American State for friendly intervention in case of European encroachment.

We have not interfered and shall not interfere, with the existing possession of any European power in this hemisphere, and to the ultimate union of all the English speaking parts of the continent by the free consent of its inhabitants; from the hour of achieving their own independence the people of the United States have regarded with sympathy the struggles of other American peoples to free themselves from European domination. We watch with deep and abiding interest the heroic battles of the Cuban patriots against cruelty and oppression, and best hopes go out for the full success of their determined contest for liberty. The government of Spain, having lost control of Cuba, and being unable to protect the property or lives of resident American citizens, or to comply with its Treaty obligations, we believe that the government of the United States should actively use its influence and good offices to restore peace and give independence to the Island.

The peace and security of the Republic and the maintenance of its rightful influence among the nations of the earth demand a naval power commensurate with its position and responsibilities. We, therefore, favor the continued enlargement of the navy, and a complete system of harbor and sea-coast defenses.

For the protection of the equality of our American citizenship

and of the wages of our workingmen, against the fatal competition of low priced labor, we demand that the immigration laws
be thoroughly enforced, and so extended as to exclude from en-
trance to the United States those who can neither read nor write.

The civil service law was placed on the statute books by the
Republican party which has always sustained it, and we renew
our repeated declarations that it shall be thoroughly and heartily,
and honestly enforced, and extended wherever practicable.

We demand that every citizen of the United States shall be allowed to cast one free and unrestricted ballot, and that such ballot shall be counted and returned as cast.

We proclaim our unqualified condemnation of the uncivilized
and preposterous [barbarous] practice well known as lynching,
and the killing of human beings suspected or charged with crime
without process of law.

We favor the creation of a National Board of Arbitration to
settle and adjust differences which may arise between employers
and employed engaged in inter-State commerce.

We believe in an immediate return to the free homestead policy of the Republican party, and urge the passage by Congress of
a satisfactory free homestead measure which has already passed
the House, and is now pending in the senate.

We favor the admission of the remaining Territories at the
earliest practicable date having due regard to the interests of the
people of the Territories and of the United States. And the Federal officers appointed for the Territories should be selected from
the *bona-fide* residents thereof, and the right of self-government
should be accorded them as far as practicable.

We believe that the citizens of Alaska should have representation in the Congress of the United States, to the end that needful
legislation may be intelligently enacted.

We sympathize fully with all legitimate efforts to lessen and
prevent the evils of intemperance and promote morality. The Republican party is mindful of the rights and interests of women,
and believes that they should be accorded equal opportunities,
equal pay for equal work, and protection to the home. We favor
the admission of women to wider spheres of usefulness and welcome their co-operation in rescuing the country from Democratic
and Populist mismanagement and misrule.

Such are the principles and policies of the Republican party.

By these principles we will apply it to those policies and put them into execution. We rely on the faithful and considerate judgment of the American people, confident alike of the history of our great party and in the justice of our cause, and we present our platform and our candidates in the full assurance that their selection will bring victory to the Republican party, and prosperity to the people of the United States.

16   FROM   *Henry Demarest Lloyd*
                    The Populists at St. Louis

Henry Demarest Lloyd had long been associated with reform and antimonopoly forces. His book, Wealth against Commonwealth, *published in 1894, was widely read and is often hailed as the first great "muckraking" book. This account of the Populist convention is tinged with bitterness. Free silver, he would later charge, was the cowbird of the reform movement. See The* Dictionary of American Biography *and Robert F. Durden, "The Cowbird Grounded: The Populist Nomination of Bryan and Tom Watson in 1896,"* Mississippi Valley Historical Review, 50:*397–423 (December 1963).*

The People's Party has "shot the chutes" of fusion and landed in the deep waters of Democracy as the Independent Republican movement of 1872 did. Nearly all the reform parties of the last generation have had the same fate. Democracy is that bourne from which no reform party returns—as yet. The Independent Republicans organized as a protest against corruption in the administration of the national government and to secure tariff reform on free trade lines. Unlike the People's Party, theirs began its career under the leadership of some of the most distinguished men in the nation. Among them were Hon. David A. Wells, who

SOURCE. *Review of Reviews, 14*:298–303 (September 1896).

had been United States Commissioner of Internal Revenue; Ex-Governor Hoadley of Ohio; E. L. Godkin, editor of the New York *Nation;* Horace White, then of the Chicago *Tribune;* Ex-Governor Randolph of New Jersey; the Hon. J. D. Cox, who had been Secretary of the Interior; Edward Atkinson of Boston; the Hon. Carl Schurz. It was the expectation of most of these gentlemen and their followers that the Cincinnati convention would nominate Charles Francis Adams of Massachusetts, our great War Minister at the Court of St. James, for President, and that with his election and a Congress pledged to civil service reform and revenue tariff the country would enter upon a new era of purity and prosperity. The revulsion when their free trade egg hatched out Horace Greeley was comparable only to that of the gold and machine Democrats at Chicago at the nomination of Bryan and the adoption of the anti-Cleveland and pro-silver platform. The People's Party had no men of national prestige to give its birth *éclat.* It has been from the beginning what its name implies—a party of the people.

One of the principal sources was the Farmers' Alliance. To President Polk of that body more than to any other single individual it owes its existence. The agrarian element has been predominant throughout its career. One of its best representatives in this convention was the temporary chairman—the Hon. Marion Butler, the handsome young farmer of North Carolina. Too young to be a candidate for President or Vice-President, he has worked his way up from his fields through the Farmers' Alliance into a seat in the United States Senate. But in addition to the revolting agrarians, nearly every other reform force—except the Socialists—has been swept into it. Its first national convention of 1892 was attended by veterans of the old Greenback movement like General James B. Weaver, by rotten-egging whom, in the campaign that followed, the Southern Democrats made tens of thousands of Populists; by anti-monopolists like Ignatius Donnelly, whose Shakespeare cryptogram has made him one of the best known writers of his day; by leaders like Powderly. It was no easy thing to find common ground for men so dissimilar to meet upon. The delicate work of preparing a platform was accomplished, thanks mainly to the skillful pen of Ignatius Donnelly. The convention went wild with joy when it became known that the Committee on Platform had succeeded in coming to an agreement and unification was assured. For over an hour the thousand members of the convention sang, cheered, danced and gave thanks. It was one of the most thrilling scenes in the panorama of American political conventions. Singularly enough, it was in the Democratic convention, this year, not that of the People's Party, that the forces of enthusiasm and revolutionary fervor flamed the brightest.

The Populist gathering of this year lacked the drill and distinction and wealth of the Republican convention held the month before in the same building. It had not the ebullient aggressiveness of the revolutionary Democratic assembly at Chicago, nor the brilliant drivers who rode the storm there. Every one commented on the number of gray heads—heads many of them grown white in previous independent party movements. The delegates were poor men. One of the "smart" reporters of the cosmopolitan press dilated with the wit of the boulevardier upon finding some of them sitting with their shoes off—to rest their feet and save their shoes, as they confessed to him. Perhaps even his merry pen would have withheld its shafts if he had realized that these delegates had probably had to walk many weary miles to get to the convention, and that they had done their political duty at such sacrifice only for conscience sake. Cases are well known of delegates who walked because too poor to pay their railroad fare. It was one day discovered that certain members of one of the most important delegations were actually suffering for food. They had had no regular sleeping place, having had to save what money they had for their nickel meals at the lunch counter. The unexpected length of the proceedings had exhausted their little store of money. Among these men, who were heroically enduring without complaint such hardships in order to attend to political duties which so many of those who laugh at them think beneath their notice, were some of the blacklisted members of the American Railway Union. They were there in the hope that they might have the opportunity of helping to make their leader, Eugene V. Debs, a candidate for President. But Mr. Debs, though he had a large following, refused to allow his name to be put before the convention, urging that every one should unite in favor of Bryan, as there seemed a chance of his election, and through him the people might at least hold their ground until ready for a more decisive advance.

In the South, the Democracy represents the classes, the People's Party, the masses. The most eloquent speeches made were those of whites and blacks explaining to the convention what the rule of the Democrats meant in the South. A delegate from Georgia, a coal-black negro, told how the People's Party alone gave full fellowship to his race, when it had been abandoned by the Republicans and cheated and betrayed by the Democrats. It was to this recognition of the colored men a distinguished political manager referred when he said recently in an interview that the Populists of the South could go where they belonged—"with the negroes." With thrilling passion the white Populists of the South pleaded that the convention should not leave them to the tender mercies of the Democrats, by accepting the Democratic nominees without the pledges or conditions which would save the Populists from going under the chariot wheels of southern Democracy. "Cyclone" Davis, spokesman of the Texas delegation, tall and thin as a southern pine, with eyes kindled with the fire of the prophet, a voice of far reach and pathos, and a vocabulary almost every other word of which seemed drawn from the Gospels or the denunciatory Psalms, wrestled and prayed with the convention to save the Populists of Texas from the fate that awaited them if they were sent back, unprotected, to their old enemies. The Democrats, the "classes," hate with a hatred like that of the Old Régime of France for the Sans Culottes of St. Antoine the new people who have dared to question the immemorial supremacy of their aristocratic rule, and who have put into actual association, as not even the Republicans have done, political brotherhood with the despised negro. This is the secret of the bolt of the Texas Populists, just announced. They have gone over to gold with the sound money men of both the old parties, because more than silver, more than anti-monopoly, the issue with them is the elementary right to political manhood. The issue in many parts of the South is even more elementary—the right to life itself, so bitter is the feeling of the old Democracy against these upstarts from the despised masses of the whites. The line between the old Democracy and Populism in the South is largely a line of bloody graves. When the convention decided to indorse Bryan without asking for any pledge from the Democrats for the protection of the southern Populists one of its most distinguished members, a member of Congress, well known throughout the country, turned to me and said: "This may cost me my life. I can return home only at that risk. The feeling of the Democracy against us is one of murderous hate. I have been shot at many times. Grand juries will not indict our assailants. Courts give us no protection."

The People's Party convention was dated to follow the conventions of the two other parties by its managers in the pessimistic belief that the Democratic party as well as the Republican would be under the thumb of the trusts and the "gold bugs." The People's Party would then have the easy task of gathering into its ranks the bolting silver and anti-monopolist Republicans and Democrats, and increasing its two millions of votes to the five and a half millions that would put it in possession of the White House for four years. It was a simple plan. That its lead would be taken from it by one of the old parties, least of all that this would be done by the party of President Cleveland and Secretary Olney, those in charge of the People's Party did not dream. The Democracy had not forgotten how they were forced to accept Horace Greeley in 1872, because the Independent Republicans had had their convention first. Its progressive elements with a leader of surpassing shrewdness and dash, Altgeld, who unites a William Lloyd Garrison's love of justice with the political astuteness of a Zach Chandler or a Samuel J. Tilden, took advantage of the tactical error of the People's Party managers in postponing its convention. The delegates as they betook themselves to St. Louis thought they saw a most promising resemblance between the prospects of the People's Party in 1896 and those of the Republican party in 1856. The by-elections since 1892 showed that its membership roll was rising and was well on the way to two millions. It was the party whose position was the most advanced on the question of social control of privileged social power, which, if contemporary literature is any guide, is the question of the times. But as the end of four years' work since the young party startled the old politicians in 1892 by showing up over a million votes in its first presidential election, the party is going this year to vote for President for one who is willing to take its votes but not its nomination. He will be its nominee but not its candidate. Such are the perplexities of the situation that it is even extremely doubtful whether the nominee will receive an official notification of his nomination or a request that he will consent to be a candidate. It is urged by influential members of the party that as a Democrat he would be "embarrassed" by such a notification and request, and that the "crisis" is so grave that they must sacrifice their party to their patriotism, and save their country by voting for the Democratic candidate without his knowledge "officially"—on the sly, as it were. Until their convention met these millions had hoped that theirs would be the main body of a victorious army. This hope ends in their reduction to the position of an irregular force of guerillas fighting outside the regular ranks, the fruit of the victory, if won, to be appropriated by a general who would not recognize them. Even more interesting is it that this is cheerfully accepted by most of the rank and file of the People's Party. No protest of sufficient importance to cause a halt was made at the first, when the shock was greatest, and the noise of dissent has grown fainter as the excitement of the campaign rises. The party is composed altogether of men who had already had the self-discipline of giving up party for the sake of principle. Every one in it had been originally either a Democrat or a Republican, and had severed all his old political ties to unite with those who, like himself, cared more for reform than old party comforts. To men who had already made one such sacrifice, another was not difficult. The People's Party is bivertebrate as well as bimetallic. It was built up of the old Greenback and Anti-monopoly, elements, reinforced by castaways of the Union Labor, National, and other third party enterprises. Its members had become well acquainted with the adversities of fusion and amalgamation, and used to being "traded" out of existence.

One of the plainest looks on the face of the St. Louis convention was anxiety—anxiety of the managers who for years had been planning to get by fusion—with Republicans or Democrats —the substance if not the name of victory, and saw in the gathering many resolute "middle of the road" opponents; anxiety of the mass of the delegates lest they were being sold out; anxiety, most surprising of all, among the radicals, lest by insisting too much upon their own radicalism they might explode a coalescence which, if left to gather headway, might later be invaluable to them. The predominant anxiety found its most striking expression in the preparation and adoption of the platform. In the committee room every suggestion for the utterance of any novelty in principle or application was ruthlessly put down. When the platform was reported to the convention, the previous question was at once moved, and the platform adopted without a word of debate. Even in the Democratic convention half a day was given to discussing the articles of political faith. No motion to reconsider this closure and secure a discussion of the principles of the movement was made! Even the radicals sat silent. In the proceedings of the convention the creed of the party was therefore practically not considered. In a large view the only subject which engrossed the gathering was whether the party should keep on in its own path or merge for this campaign with the Democracy. The solicitude to do nothing which should hinder the Rising of the People, if that had really begun, was the motive that led to the indorsement of Bryan. Most of the three hundred, over one hundred of them from Texas alone, who refused to unite in this, would have joined its one thousand supporters had the protection they prayed for against the old Democracy been given them by the exaction of guarantees from the Democratic candidate and campaign managers. It was not that they loved Bryan less. A determination that the People's Party and that for which it stood should not be lost if this year's battle was lost by its ally, Democracy, accounts for the nomination of Watson. The majority which insisted that all the precedents should be violated and the Vice-President nominated before the President, and which rejected Sewall and took Thomas E. Watson of Georgia—a second Alexander H. Stephens in delicacy of physique and robustness of eloquence and loyalty to the people—was composed, as the result showed, mostly of the same men who afterward joined in the nomination of Bryan. It is true there was a strong opposition to Sewall, because he was national bank president, railroad director and corporation man. But the nomination speeches and the talk of the delegates showed convincingly that the same men who meant to support Bryan were equally well minded that there should not be an absolute surrender to the Democracy. The Democracy must yield something in return for the much greater concession the People's Party was to give.

Contrary to expectation and to the plan by which the two conventions had been brought to St. Louis on the same dates, the silver convention exercised no influence on that of the Populists. The delegates of the latter listened with unconcealed impatience to every reference to the silver body, and refused to allow its members any rights upon the floor. The report of the Conference Committee was listened to without interest. The tumultuous refusal of the convention to allow Senator Stewart of the silver convention an extension of time when he was addressing them, was one of the many signs that the convention cared less for silver than did the Democratic convention. Most of the Democrats really believe free silver is a great reform. That is as far as they have got. But it was hard to find among the Populists any who would not privately admit that they knew silver was only the most trifling installment of reform, and many—a great many— did not conceal their belief that it was no reform at all. The members of the People's Party have had most of their education on the money question from the Greenbackers among them— men like the only candidate who contended with Bryan for the nomination before the convention—Colonel S. F. Norton, author of the "Ten Men of Money Island," of which hundreds of thousands of copies have been sold, who for twenty years has been giving his means and his life energy to agitating for an ideal currency. The People's Party believes really in a currency redeemable in all the products of human labor, and not in gold alone, nor in gold and silver. A party which hates Democracy accepted the Democratic nominee, and a party which has no faith in silver as a panacea accepted silver practically as the sole issue of the campaign. Peter Cooper, the venerable philanthropist, candidate for President on the Greenback ticket in 1876—whose never absent air cushion Nast by one of his finest strokes of caricature converted into a crown for General Butler when running as Greenback and Labor candidate for Governor of Massachusetts— presided over the first days of the convention from within the frame of a very poorly painted portrait. But later, by accident or design, about the time when it thus became plain that the convention would make only a platonic declaration of its paper money doctrines, and would put forward only "Free Silver" for actual campaign use, the face of the old leader disappeared and was seen no more with its homely inspiration above the chairman's head.

The solution of the paradoxical action of the convention as to Democracy and money was the craving for a union of reform forces which burned with all the fires of hope and fear in the breasts of the delegates, and overcame all their academic differences of economic doctrine and all their old political prejudices. The radicals had men who were eager to raise the convention against the stultification they thought it was perpetrating. If the issue had been made there was an even chance, good arithmeticians among the observers thought, that the convention could have been carried by them, and a "stalwart" ticket put into the field on a platform far in advance of that adopted in Omaha in 1892, one demanding, for instance, the public ownership of all monopolies. This contingent felt that the social question is more than the money question, the money question more than the silver question, and the silver question more than the candidacy of any one person. If the money question was to be the issue it wanted it to be the whole money question—the question how an honest dollar can be made instead of being only stumbled on in placers or bonanzas, and how it can be made as elastic as the creative will of the people and as expansive as civilization itself. Certainly the strongest single body of believers in the convention was this of anti-monopoly in everything, including the currency. These men would much rather have declared for the demonetization of gold than the remonetization of silver. That their strength was formidable—formidable enough to have split the convention near the middle, if not to have carried it—no one could deny who studied on the ground the feelings and beliefs of the delegates. But those who might have called this force into activity were quiescent, for Col. Norton's candidacy was unsought, impromptu and without organization. The leaders did not lead, and their followers did not clamor to be led. "General" J. S. Coxey of the Commonweal Army, who has left large property interests to suffer while he has devoted himself to educating the people on his "Good Roads" plan of internal improvements, to be paid for by non-interest bearing bonds, was present, and made no resist-

ance outside of the Committee of Resolutions. Ex-Governor
Waite of Colorado, whose name will be cheered in any assembly
of labor men or Populists, as the only Governor who has called
out the militia to protect the workingmen against violence at the
hands of their employers, for the sake of harmony forbore to
press his claims at the head of a contesting delegation from Colorado. Senator Peffer, who has shown an ample courage in every
emergency at Washington, sat silent, though he was bitterly opposed to the methods of the managers. The fear ruled that unless
the reform forces united this time they would never again have
the opportunity to unite. It was in the air that there must be union. The footfall of the hour for action was heard approaching. It
was a phsychological moment of *rapprochement* against an ap-
palling danger which for thirty years now had been seen rising in
the sky. If the radicals made a mistake, it was a patriotic mistake. The delegates knew perfectly well that the silver miners
were spending a great deal of money and politics to get them to
do just what they were doing. They knew what the Democratic
politicians were doing with the same object. They knew that with
some of their own politicians the anxiety to return to the old political home was not dissociated from visions of possible fatted
calf. But though they knew all this, they went on by an overwhelming majority to do what the mine owners and the Democrats and the traders wanted them to do, and the acquiescence of
the mass of the party in their action is now beyond question. We
can comprehend this better when we see men like Edward Bellamy, the head of the Nationalists, and Henry George of the Single
Taxers, and the Rev. W. D. P. Bliss of the Christian Socialists
also taking the same attitude and for precisely the same reason
that the real issue is "between men and money," in Bellamy's
phrase; and they cannot afford to side with money against men.

17 FROM   *Newell D. Hillis*
An Outlook upon the Agrarian Propaganda in the West

Henry Demarest Lloyd lamented the Populists' espousal of the silver issue because he believed it subordinated important reforms that should be kept in view. But Newell Dwight Hillis, a Republican, a Presbyterian minister, a Chautauqua lecturer, and an organizer of Sunday schools found, or seemed to find, the free silver advocates a real danger to the Republic.

Recent discussions and editorials in the various journals and reviews of New York seem to indicate that the East does not fully understand either the strength of the silver sentiment or the methods and arguments by which it is being advanced in the interior and West. During several weeks past I have been lecturing before various Chautauquas, summer assemblies and colleges of Ohio, Illinois, Indiana, Michigan, Wisconsin and Iowa. These summer assemblies, continuing through ten or twelve days with their summer schools, lectures and concerts by the best platform speakers of the country, assemble audiences at once vast and widely representative. Here pulses and throbs the intellectual life of the entire section. Conversation with a large number of representative men has convinced me that as Republicans we must adopt new methods of discussion and redouble our energies if we are to destroy the silver heresy and maintain sound money. The outline of a single address given to an assembly of farmers in a country schoolhouse in Iowa will interpret the methods and arguments used throughout the entire West.

The chief feature of the speaker's address was his charts. Upon one end of a blackboard was written an estimate of the

SOURCE. *Review of Reviews, 14:*304–305 (September 1896).

number of millions of bushels of oats raised this year by the farmers of Iowa, and a further estimate of the value of the crop at the market price of 13 cents a bushel. The Populist portrayed the farmer working like a slave through eight months of the year to produce this 13-cent bushel of oats, while the railway in a single day and night hauled the grain to Chicago, where it receives 7 of the 13 cents as its recompense. Now the first cent of the seven extorted will, urged the orator, take away all hope of the farmer paying the interest on his mortgage; the second cent will take from wife and daughter woolen dress warm against the winter; the third will take the boy and girl out of school and college and condemn them to the drudgery of the farmhand or housemaid; the fourth cent will take away all possibility of purchasing the review, the newspaper, the book, and drive men back to barbarism. When the orator reached this point in his discussion the audience was inflamed to the highest point. At that moment self-interest and prejudice armed his listeners against all arguments for sound money. Had the Republican committee been there when the assembly dispersed to present each farmer with a library devoted to the exposure of the silver heresy, even the multitude of books would not have availed for reversing the farmer's judgment or convincing him that the gold standard is not responsible for his misfortunes, or that free silver is not the unfailing panacea for all his ills.

In many of the rural districts class hatred and sectionalism are invoked against McKinley and the Republican party. The farmer is told that the reason why the railroads extort 7 cents out of the 13 paid for the bushel of oats is that the railroad must pay interest on watered stock representing two or three times the cost of building the road. Now the argument of the Populist is that this water must be squeezed out of the stock before the farmer can hope for better rates. As a means to this desired end it is urged that since railways cannot increase the fare of three cents a mile, the success of free silver will throw the railway into the hands of a receiver and force an entire readjustment. Like dynamite, class hatred is a powerful weapon, and the farmer is urged to use it against his ancient enemy, the corporation. By the skillful use of the half truths and falsehoods the prophet of free silver succeeds in inciting the farmer to punish the railways in the hope that some time in the long run benefit will accrue to him in the shape of lessened charges for transportation.

Strangely enough, one of the most effective arguments that is being used is directed not against capital, nor against ability as represented by the employer, but against the trades unions of the cities. The farmers affirm that carpenters, plasterers and masons have, through strikes and riots, succeeded in maintaining a false standard of wages. In the face of the falling prices for the farmer, with wheat selling for 60 cents a bushel, the carpenter and mason has, through the long period of financial depression since 1893, held his wage up to 40 and 50 cents an hour, all this, too, despite the fact that the farmer of the great interior and western states has during the same period toiled not eight hours a day, but fourteen or sixteen, and received on an average but 78 cents per day. By reason of their isolation the farmers feel that it has been impossible for them to organize trades unions enabling them to maintain their rights in the same way that the laboring men in the cities have defended themselves against wrong. Now the problem that fronts the farmer, the Populist urges, is how shall the wage of the laborer in the city be equalized with the wage of the laborer in the pasture or meadow. In nature there is a law by which the water in the spout of the tea-kettle finds the same level with the water in the kettle itself. But wages will not equalize themselves; the task of equalization asks the farmer's aid. The gist of the silver orator's argument touching this point is this: Suppose Bryan is elected and the country goes to a silver basis. The carpenter's or mason's wages will still stand at 40 or 50 cents an hour, for at the very best he can scarcely hope for an advance in wages of more than 5 or 10 cents an hour. But with the small increase in amount of wages will come the halving of the purchasing power of his money. But for his 60-cent bushel of wheat the farmer will, under the new conditions, obtain $1.20. Not capital, not ability, not labor, but land, therefore, is to receive the benefit of the financial change. Thus the wages of the farmer will be made to approach those of the carpenter or mason, and that, too, without riot or strike or the use of arms.

Unfortunately this method secures the transfer of a part of the wages from the pocket of the carpenter or mason in the city to the pocket of the farmer in the country. It gains for one class of workingmen at the expense of another. It is my firm conviction that the election of McKinley and the success of the principles, financial and economic, for which he stands, will increase the farmer's wage without lessening the wage of the laboring men in cities. A box filled with ballots representing such arguments and half-truths would not equal a single vote cast by wise men in the days of Adams, Hamilton and Jefferson.

Much is being said about the campaign of education. Unfortunately, unto the present moment the education has been largely on the part of the Populists. The zeal of the silver orator is something to stir the wonder and alarm of all intelligent men. Like the zealot of old, the silverite rises yet a great while before day to compass one convert before milking his cows or finding his way into the fields. All day long he hastens his footsteps that he may have an hour in the evening for visiting some unconvinced neighbor. He returns from the field to take up the argument where he dropped the thread in the morning. He counts himself the divinely ordained apostle of the new financial movement. He goes to church on Sunday to obtain inspiration for prosecuting his mission during the week. Farmers' picnics by streams and in groves are held. The bicycle race, the horse race, the wrestling match and the silver debate increase the crowds. When the sound money orator begins his argument he finds himself working against signal odds. He who starts out to convert others finds it hard to confess he himself has been wrong. He is impervious to argument. His mind may be compared to a bottle empty and corked as it floats in the sea. The ocean itself cannot fill such a bottle, and the larger the ocean and the greater the vacuum of the bottle, the tighter is the cork pushed in. Under such conditions the old orthodox methods of campaign are impotent. A new kind of literature even must be evolved. Many difficulties hitherto unknown have been developed.

Then the successful tariff speaker is not always a successful disputant of the financial question. A clear view of the silver question involves wide reading and experience and a trained mind,—conditions asking for years, not weeks of education. Up to the present moment the great need in the Republican campaign is a need of illustrated literature. A short, spicy statement with a cartoon or picture will distribute itself; it has wings and feet and walks or flies throughout the township or county. Contrarywise, long pamphlets, studied financial discussions and the abstract documents sent out will never be read by farmers, but will serve during the coming winter for lighting the kitchen fire of the man who is supposed to distribute them. One of the members of the English Cabinet has said that Lord Rosebery was defeated and Salisbury elected by reason of the large posters pasted on barns and the cartoons sent out through patent insides of newspapers. Beyond a peradventure, a new kind of campaign document must be invented. The eye offers a short route to reason and judgment. The poster as an influence in the campaign offers more hope than any other method of public instruction.

After patient investigation I am convinced that the present industrial depression has its explanation in causes other than the appreciation of gold or the depreciation of silver. In the long run the farmers not less than the laboring men in cities have only misfortune and sorrow as the result of the election of Bryan. But my acquaintance with the rural districts of states like Illinois and Iowa makes it impossible for me to believe that the farmers will ever consent to a policy of repudiation. These states were settled largely by New England in connection with the Kansas and Nebraska troubles in 1857. No section in the entire country represents a higher average of intelligence and culture; no section buys more books and magazines, or sends a larger proportion of its young men and women to the academy and college. Beecher and Gough used to say no section in the land gave a more appreciative hearing. The country district has always furnished the leaders to the city. Eighty five per cent of the great financiers, lawyers, bankers, merchants and professional men of the cities have come from the country, or from the small villages. The leaders of the next generation in the city are to-day toiling behind the plow in the country. I have abiding confidence in the intelligence and morality and sober second thought of the farmers and their sons. Once the question is fully before them they will refuse dishonor and repudiation.

18 FROM *Edwin Markham*
The Man with the Hoe

Edwin Markham's "Man with the Hoe" virtually exploded across the United States in 1899. Louis Filler in his essay on Markham in the Dictionary of American Biography *(supplement 2) says that it was translated into 40 languages and was reported to have earned its author a quarter of a million dollars in royalties. Mark Sullivan in Our Times (2:239ff.) describes its impact. It may well be the only poem ever published in America to be discussed and belabored in labor councils and grange halls, in churches, schools, colleges, and corporation boardrooms, and by politicians of almost every stripe. Mark Sullivan says that Collis P. Huntington, a railroad millionaire, offered a $700 prize anonymously for the best poetical answer to it. In publishing the poem, Markham asserted that it had been inspired by Millet's* Man with the Hoe, *but to millions of Americans who had never seen Millet's painting it probably spoke to the ruins of the dream of what America had promised. William Jennings Bryan, the defeated presidential candidate of the Populists and Democrats, like the poem. He said it "voices humanity's protest against inhuman greed" (quoted in Our Times, 2:243).*

(Written after Seeing Millet's World-Famous Painting)

"God made man in His own image,
    in the image of God made He him."

Genesis

Bowed by the weight of centuries he leans
Upon his hoe and gazes on the ground,

SOURCE. Edwin Markham, *The Man with the Hoe and Other Poems*, Doubleday and McClure, New York, 1899, pp. 15–18. © Edwin Markham and *The San Francisco Examiner.*

The emptiness of ages in his face.
And on his back the burden of the world.
Who made him dead to rapture and despair,
A thing that grieves not and that never hopes,
Stolid and stunned, a brother to the ox?
Who loosened and let down this brutal jaw?
Whose was the hand that slanted back this brow?
Whose breath blew out the light within his brain?
Is this the Thing the Lord God made and gave
To have dominion over sea and land;
To trace the stars and search the heavens for power;
To feel the passion of Eternity?
Is this the dream He dreamed who shaped the suns
And pillared the blue firmament with light?
Down all the stretch of Hell to its last gulf
There is no shape more terrible than this—
More tongued with censure of the world's blind greed—
More filled with signs and portents for the soul—
More fraught with menace to the universe.

What gulfs between him and the seraphim!
Slave of the wheel of labor, what to him
Are Plato and the swing of Pleiades?
What the long reaches of the peaks of song,
The rift of dawn, the reddening of the rose?
Through this dread shape the suffering ages look;
Time's tragedy is in that aching stoop;

Through this dread shape humanity betrayed,
Plundered, profaned and disinherited,
Cries protest to the Judges of the World,
A protest that is also prophecy.

O masters, lords and rulers in all lands,
Is this the handiwork you give to God,
This monstrous thing distorted and soul-quenched?
How will you ever straighten up this shape;
Touch it again with immortality;
Give back the upward looking and the light;
Rebuild in it the music and the dream;

Make right the immemorial infamies,
Perfidious wrongs, immedicable woes?

O masters, lords and rulers in all lands,
How will the Future reckon with this Man?
How answer his brute question in that hour
When whirlwinds and rebellion shake the world?
How will it be with kingdoms and with kings—
With those who shaped him to the thing he is—
When this dumb Terror shall reply to God
After the silence of the centuries?

19 FROM *William V. Allen*
The Populist Program, 1900

By 1900 the strength of the Alliance Movement had waned. A successful war with Spain, the return of some prosperity, improved weather in the wheat country, and a host of other developments had changed the mood of the country. Many of the Populists, the farmers and their allies, had returned to the Democratic or the Republican party. This modest statement of the Populist program of 1900 by William V. Allen, Populist Senator from Nebraska, 1893 to 1899, suggests that the fires of reform no longer burned brightly.

The essential elements of the Populist platform will be the same as in 1896. In regard to silver the position of the party has not been and will not be changed. The whole money question is as live an issue now as it ever was with the people of the West. The failure to agitate it as it was agitated in 1896 is due to the

SOURCE. *Independent,* 52:475–476 (February 1900).

fact that the last Presidential campaign is well by and the next
one has not yet begun. It has ceased to be a matter of controversy with our people; it has passed into a conviction, like, a man's
conviction on religion, or on any other subject which to his mind
is settled. The money question does not stop with the silver question; that is only one feature of it. The Populist party is in favor
of withdrawing all issue power from the national banks and having all money, gold, silver and paper, issued directly by the Government. We hold that that is a constitutional prerogative which
should be exercised in the interest of our own people. To turn
over that power to the national banks is to abandon the power
that was conferred on Congress for the benefit of the nation at
large. It is class legislation and puts in the hands of a few thousand institutions the dangerous power of contracting and expanding the currency and by that means taking from the people
much of their property and earnings, to say nothing of the ab-
surdity of permitting corporations to use their debts as money
and forcing people to take that money. We want all forms of
money to be legal tender; there should be no qualification. The
power to enforce that tender should reside in the courts of law.
Legal tender has been the right of the debtor for centuries, until
changed by act of Congress.

The trust question is bound to be an issue in the next campaign. It is not a new question with the Populists. They were pioneers in the silver question and early took a stand against trusts.
National legislation will be demanded along the lines of the national platform. While other parties are divided on this matter
the Populist party is arrayed solidly against trusts.

There is a marked distinction between expansion and imperialism. Expansion is a natural, orderly, national growth and the
acquisition of territory for our own population with a view of ultimately making the territory acquired a State or States, and the
inhabitants thereof citizens of the United States. Imperialism, as
I understand it, is simply the Napoleonic method of forcibly acquiring remote lands and peoples with a view of spoliation and
aggrandizement and with no view of making the acquired territory States of the union or the people citizens of the United States.

It involves a large standing army, large enough to increase
taxation at home and to impose other home burdens. Necessarily

it will draw us into complications with European and Asiatic politics, which we should avoid for our safety. I am speaking for myself upon this matter. The traditions of this Government are against a large standing army. It is a menace to the independence of the people, an unnecessary burden to the taxpayer, and is not consonant with the spirit of free institutions. Our party, I think, views the question in that light.

The Populist party on general principles is arrayed against an excessive tariff.

In regard to the Nicaragua canal, our party would be against the issue of bonds and subvention but in favor of acquiring the canal and paying for it out of the revenues of the Government as the work goes on, not leaving a debt to be paid by another generation.

We are interested in the extension of commerce. The party on general principles is opposed to the policy of subsidies and bounties. It is not opposed to legitimate and well-devised Government aid to new enterprises that bid fair to become of importance and value to the people; but it is opposed to continuing subsidies and bounties to enterprises which have demonstrated their ability to care for themselves. I am rather inclined to think that the party would favor discriminating duties in favor of American vessels, reaching the end in that way.

PART FOUR

The View from the Study

Much has been written about the Granger Movement. In 1873, two years before the Patrons of Husbandry reached their largest membership, James Dabney McCabe published his *History of the Grange Movement; or, the Farmer's War Against Monopolies. . . .* under the pseudonym of Edward Winslow Martin; in 1874, Jonathan Periam's *The Groundswell: A History of the Origin, Aims, and Progress of the Farmers' Movement. . . .* appeared, and the next year Oliver Hudson Kelley published his account of the rise of the Patrons of Husbandry under the title, *Origins and Progress of the Order of the Patrons of Husbandry in the United States; a History from 1866 to 1873.* Solon J. Buck's scholarly account, *The Granger Movement: A Study of Agricultural Organization and its Political, Economic and Social Manifestations, 1870–1900,* was published in 1913. The extent to which the Granger Movement has attracted the attention of historians is reflected in a 21-page bibliography prepared by Dennis S. Nordon in 1967, *A Preliminary List of References for the Granger Movement.* In his book, *Railroads and the Granger Laws,* published in 1971, George H. Miller does a masterly job of showing the part of the Patrons of Husbandry, the Grangers, and other forces played in bringing about the railroad rate laws in the 1870s.

The Greenback Movement, which touched some of the farmer reform interests, has been treated in detail most recently by Irwin Unger in his *The Greenback Era: A Social and Political History of American Finance.*

The Alliances and the Agricultural Wheels had their early chroniclers in W. Scott Morgan, N. A. Dunning, and a host of other contemporary defenders and detractors. John D. Hick's *The Populists Revolt: A History of the Farmers' Alliance and the People's Party,* provided the first general, scholarly account of the movement when it appeared in 1931. His view of the Populists as fairly enlightened reformers who provided much of the basis for many Progressive and New Deal reforms was widely accepted until the end of the 1940s. New concerns, however, and new questions about what America had been and what it was becoming led many to challenge this view. During the 1950s and 1960s historians and others reexamined the story of the Populists. Some discovered, or thought they had discovered, dark and ugly aspects of Populism—anti-Semitism, other racism, and religious bigotry—not perceived by Hicks and his followers. That the revisionists made some converts to their view is illustrated by the casual observation by Richard Rovere, a political reporter for *The New Yorker,* about the "Populists" of 1972 who differs from "their predecessors mainly in having shed the racial and religious bigotry that once was thought of as part of the doctrine" (July 22, 1972, page 72).

The three articles that follow are benchmarks in the continuing reexamination of the character and meaning of the Populist Movement. Professor Hicks' article, published in 1949, is concerned primarily with tracing the spirit of reform and reforms that flowed from the Populists' campaign. By the time Professor Woodward's article was published in 1959 the Populists had been under severe attack from many quarters as the source of several ugly aspects of American society. Although not denying the validity of some of the charges, Professor Woodward seeks to bring discussion back to the consideration of who the Populists were and what they sought. By 1965, when Professor Hanlin's article was published, many of the revisionists charges had been challenged, but Professor Hanlin complained that the result had been mostly a sterile argument. He proposed approaches that would lead to a fuller, better understanding of the Populist Movement.

The last two articles indicate something of the intensity of the discussion by historians and others of the character and program of the Populists. Perhaps it is not unfair to add that they also reflect the extent to which the basis for farmer discontent and desperation during this period of great growth, change, and dislocation of farming has been submerged in discussion of the political reforms that farmer organizations had helped to launch.

20 FROM *John D. Hicks*
The Legacy of Populism in the Western Middle West

Ever since President Conant of Harvard University voiced the opinion some years ago that one of the things the United States needed most was a rebirth of American radicalism to offset the alien varieties being currently imported, historians have been at pains to try to set the noted educator right.[1] American radicalism, most of them contend, is by no means dead, but continues as one of the most active aspects of the American way of life. Its roots lie deep, and its trunk stands strong. It draws its strength from a long line of liberal humanitarians, from a long line of laborites, from a long line of western agrarians. Each of these three groups has influenced the others, and all have drawn on European experience and ideas. But the resulting brand of radicalism is fundamentally an American, not a European, product. And it still endures.[2]

We are at this time primarily concerned with the contributions that nineteenth-century agrarians made to the later radicalism of what is sometimes called the western Middle West—or as the census takers would say, with awkward accuracy, the West North Central States, plus Wisconsin and Illinois. It was in this region that the farmers' voice of protest was most eloquently and persistently raised. It was here, during hard times, that a philosophy of radicalism was devised sufficient even to endure

SOURCE. *Agricultural History, 23*:225–236 (October 1949).

[1] This article by Professor John D. Hicks is the presidential address presented at the annual meeting of the Agricultural History Society in Washington, D. C., on Sept. 13, 1949—*Editor*.

For the views of President James B. Conant on radicalism, see his article, "Wanted: American Radicals," in *Atlantic Monthly, 171*:41–45 (May 1943).

[2] Chester McArthur Destler, "Western Radicalism, 1865–1901: Concepts and Origins," in *Mississippi Valley Historical Review, 31*:354–355 (December 1944).

the acid test of good times during the early twentieth century. What the farmers of this area had suffered from railroads, banks, middlemen, and manufacturers had made them convinced antimonopolists. They were by no means the first to hold antimonopoly views, for the ideas they expressed had often been cogently stated by eastern, or even European, theorists. But the long struggle with frontier poverty, culminating in the Populist revolt, had instilled in many farmer minds a deep-seated belief that the various combines through which big business operated must somehow be restrained. This attitude was not due to ignorance, but to experience. The farmers knew whereof they spoke. Nor did they have any doubt as to the role the government must play in providing this restraint. Middle Western agrarians were not socialists; on the contrary, they were, or at least they aspired to be, small capitalists. But their property-mindedness did not blind them to the fact that only the power of government could insure them against the unfair advantages of monopoly. They favored government regulation and control, or in extreme cases government ownership, only as a means of retaining for themselves the right to hold property and to do business on a reasonably profitable basis.[3]

It should be remembered that these rights, throughout the greater part of the western Middle West, were most imperiled by the railroads, and that it was against the railroads, more specifically than against any other type of enterprise, that the farmers aimed their principal reforms. The railroads had created the region; they had brought the population in; they were in close alliance, or even partnership, with other industries, such as lumber, elevator, milling, and packing corporations; they were the chief exploiters of the farm population which was obliged to pay them rates, both coming and going. When the average Middle Western farmer living west of Chicago talked about monopolies and

[3] Benton H. Wilcox, "An Historical Definition of Northwestern Radicalism," in *Mississippi Valley Historical Review*, 26:377–394 (December 1939). This article sets forth the principal findings of the author's more elaborate study, A Reconsideration of the Character and Economic Basis of Northwestern Radicalism, unpublished doctoral dissertation, dated 1933, in the library of the University of Wisconsin, hereafter referred to as Wilcox, Northwestern Radicalism.

trusts, he was thinking primarily of the railroads. Even the towns and cities were peculiarly railroad-conscious. They had no other equally big businesses, and their very lives as trading centers depended upon the fairness, or sometimes the favor, with which railroad rate makers treated them.[4]

Implicit in the Populistic concept of governmental intervention in economic affairs was the assumption that the government itself should be truly representative of the people, that the long-established control of the "plutocrats" should be broken.[5] The first task that the agrarian leaders set for themselves, therefore, was to capture for the people the machinery of government. It was with this end in view that Farmers' Alliance and Populist candidates sought control of State governments, and that the Populist Party nominated Weaver in 1892 and Bryan in 1896 for the presidency of the United States. Bryan's first defeat rang the death knell of Populism as an effective party organization, and served notice on the people generally that the ousting of the "plutocrats" was to be no easy task. But the idea lived on. As Frederick Jackson Turner once phrased it, "Mr. Bryan's Democracy, Mr. Debs' Socialism, and Mr. Roosevelt's Republicanism all had in common the emphasis upon the need of governmental regulation of industrial tendencies in the interest of the common man; the checking of the power of those business Titans who emerged successful out of the competitive individualism of pioneer America."[6] If this end were ever to be attained, however, the people must somehow take over their government, and the desire to see this ambition achieved survived intact long after the disappearance of the Populist Party.

Throughout the western Middle West, and to a considerable extent throughout the country as a whole, this legacy of Populism determined the course of political development during the opening years of the twentieth century. What reforms could be instituted to make sure that the people really governed? The movement for the direct primary, for the initiative and referen-

[4] Wilcox, Northwestern Radicalism, 50.
[5] John D. Hicks, *The Populist Revolt* (Minneapolis, 1931), 405–406.
[6] Frederick Jackson Turner, *The Frontier in American History* (New York, 1920), 281.

dum, and for various other aspects of popular government grew naturally out of the soil prepared by the Populists. The campaign to limit the power of the Speaker of the national House of Representatives was led by an outraged Nebraskan, George W. Norris.[7] The activities of insurgents and progressives generally, culminating in the formation of the Progressive Party of 1912, followed an evolutionary pattern closely connected with Populism. This, I repeat, is not to say that the *only* force back of twentieth-century American radicalism was nineteenth-century Middle Western agrarianism. The contributions of the liberal humanitarians and the laborites must not be overlooked, nor need the influence of imported socialistic ideas be discounted. But when all is said and done, American radicalism would simply never have been what it was but for its long and sturdy line of Granger-Greenback-Populist progenitors. As convinced antimonopolists these reformers believed that the state must use its power to regulate and control the "trusts," most of which, in the western Middle West, turned out to be railroads. They believed, too, that if the state were to be charged with this responsibility its power must be lodged firmly in the hands of the people. Probably they expected greater results from the popular rule than was reasonable, but, judged by any standards, they did accomplish a great deal. As a result of their efforts "something new had been brought into politics."[8]

• • •

The reforms inaugurated by the legislatures of the western Middle West during this period were by no means identical, but two clearly-defined objectives stood out preeminently in every State, namely, popular rule, and corporation control.

One type of legislation created, well in advance of most of the other States of the union, a direct primary system of making nominations for office. This reform was fundamental, and throughout the western Middle West it literally revolutionized State government. Nor was there in this region any such backsliding and evasion as occurred in some of the Eastern States.

[7] *Fighting Liberal: The Autobiography of George W. Norris* (New York, 1945) , 107–119.

[8] Wilcox, Northwestern Radicalism, 107.

The change had come to stay; candidates were at the mercy of
public opinion in a way they had never been before. Sometimes,
but not always, the direct primary was supplemented by plans
for direct legislation, the initiative and referendum. Honesty in
politics was frequently sought by means of drastic anti-lobbying
and corrupt practices acts. Direct primaries for candidates for
the United States Senate became common, and in some instances
a preferential vote, taken at the time of the regular election,
bound the legislature to accept the candidate designated by the
people at the polls. Presidential preference primaries for the
selection of delegations to the national nominating conventions
were also frequently provided, particularly as a result of the candidacy of Theodore Roosevelt against Taft for the Republican
nomination in 1912.[9]

Along with these efforts to promote popular government came
much legislation aimed at the political and economic supremacy
of powerful business interests, particularly the railroads. Expansion of the prerogatives of railroad commissions, higher corporation taxes, maximum freight rates, two-cent passenger fares, and
anti-pass laws were multiplied in State after State. It is no exaggeration to say that, for the most part, the peculiar hold that the
railroads had long had upon the political life of the region was
broken. Even the conservative reaction, which began in the western Middle West as early as 1912, and swept numerous "Stand-
patters" and "Stalwarts" back into office, failed to alter this picture materially. For the only way in which the conservatives
could retain their power was to outdo the Progressives in their
devotion to the new reforms. The old Populist principle that, if
the people could only obtain control of their government, they
could defend themselves adequately against the power of monopoly, seemed in process of being demonstrated.[10]

The legacy of Populism could easily be traced, also, into the
realm of national politics. Such reforms as came to be associated
with the name of Theodore Roosevelt were ardently supported
by the agrarian leaders of the western Middle West—to some extent, no doubt, were inspired by them. According to one enthusi-

[9] Merriam and Overaker, *Direct Primary*, 62–63, 41–142; Wilcox, Northwestern Radicalism, 107.
[10] *Ibid.*, 110–114; *Wallaces' Farmer*, 33:976 (Aug. 14, 1940).

ast, Roosevelt was "the spokesman of the people, the expression and exponent of the reform spirit, the mouthpiece of an awakened conscience." But the westerners were willing to go much further than Roosevelt was willing to lead. As La Follette put it, "He acted upon the maxim that half a loaf is better than no bread. I believe that half a loaf is fatal whenever it is accepted at the sacrifice of the basic principle sought to be attained."[11] The insurgent movement of the Taft administration was even more obviously of agrarian origin. It was, indeed, mainly the work of Senators and Representatives from the Middle West, men who, according to William Allen White, "caught the Populists in swimming and stole all of their clothing except the frayed underdrawers of free silver."[12]

The fight on Cannonism in the national House of Representatives was carried to a successful conclusion through the leadership of such Middle Western progressives as Norris of Nebraska, Nelson of Wisconsin, Murdock and Madison of Kansas, and Lindbergh of Minnesota. Aid came from some outside supporters, notably Poindexter of Washington and Fowler of New Jersey, but the credit for victory belonged primarily to the Middle Western agrarians.[13] The assault of the Senate insurgents upon the Payne-Aldrich tariff bill was almost wholly a contribution of the western Middle West. La Follette of Wisconsin, Clapp of Minnesota, Cummins and Dolliver of Iowa, and Bristow of Kansas were the outstanding leaders; only Beveridge of Indiana deserves comparable credit for the work the Insurgents did in revealing the monopolistic intent of the Aldrich schedules.[14] In both houses of Congress the Insurgents fought also for a graduated income tax, for conservation, for postal savings, for more vigorous railroad regulation, and against a type of reciprocity with Canada designed to benefit the industrial East at the ex-

[11] *La Follette's Autobiography*, 388; Henry F. Pringle, *Theodore Roosevelt* (New York, 1931) , 419.

[12] Kenneth W. Hechler, *Insurgency; Personalities and Politics of the Taft Era* (New York, 1940) , 21–22.

[13] *Ibid.*, 33–43.

[14] *Ibid.*, 83–91, 145.

pense of the agricultural Middle West.[15] The overwhelming approval of Middle Western farmer constituencies for the program of the Insurgents was repeatedly demonstrated at election time; not only were the radical leaders consistently returned to Congress, but old guard conservatives were retired with great good will. Eventually most of the reforms for which the Insurgents stood found expression in the platform of the Progressive Party of 1912, but the candidacy of Theodore Roosevelt blurred the issue, and divided their forces. They could not very truly believe in him, nor he in them. But the statement of a close student of the subject that "Wilsonian liberalism and the New Deal were born of Insurgency" carries no appreciable discount.[16]

[15] *Ibid.*, 146–219. Reciprocity with Canada, as proposed by the Taft Administration, was according to one Middle Westerner "a jug-handled affair," wholly unsatisfactory to the friends of genuine reciprocity. It "assumes that the farmer owes the manufacturer a living." *Wallaces' Farmer, 36*:438 (Mar. 10, 1911) .

[16] Hechler, *Insurgency,* 221.

21 FROM   *C. Vann Woodward*
The Populist Heritage and the Intellectual

During the long era of the New Deal one had little difficulty living in comparative congeniality with the Populist heritage. The two periods had much in common, and it was easy to exaggerate their similarities and natural to seek antecedents and analogies in the earlier era. Because of the common setting of severe depression and economic dislocation, Populism seemed even closer to the New Deal than did Progressivism, which had a setting of prosperity. Common to both Populists and New Dealers was an antagonism to the values of the dominant leaders of the business community bordering on alienation. They shared a sense of ur-

SOURCE. *The American Scholar, 29*:55–72 (Winter 1959–1960) .

gency and an edge of desperation about the demand for reform. And in both, so far as the South and West were concerned, agricultural problems were the most desperate, and agrarian reforms occupied the center of attention. It seemed entirely fitting that Hugo Black of Alabama and Harry Truman of Missouri—politicians whose political style and heritage were strongly Populistic —should lead New Deal reform battles. From many points of view the New Deal was neo-Populism.

The neo-Populism of the present bred a Populistic view of the past. American historiography of the 1930's and 1940's reflects a strong persuasion of this sort. The most popular college textbook in American history was written by a Midwesterner who was friendly to Populism and was himself the foremost historian of the movement. The leading competitor among textbooks shared many of the Populist leanings, even though one of its authors was a Harvard patrician and the other a Columbia urbanite. A remarkably heterogeneous assortment struck up congenial ties in the neo-Populist coalition. Small-town Southerners and big-city Northerners, Texas mavericks and Hudson River aristocrats, Chapel Hill liberals and Nashville agrarians were all able to discover some sort of identity in the heritage. The South rediscovered ties with the West, the farmer with labor. The New York-Virginia axis was revived. Jacksonians were found to have urban affiliations and origins. Not to be outdone, the Communists staked out claims to selected Populist heroes.

Many intellectuals made themselves at home in the neo-Populist coalition and embraced the Populist heritage. They had prepared the way for the affiliation in the twenties when they broke with the genteel tradition, adopted the mucker pose, and decided that conventional politics and the two major parties were the province of the boobocracy and that professional politicians were clowns or hypocrites. In the thirties intellectuals made naïve identification with farmers and workers and supported their spokesmen with enthusiasm. The Populist affinity outlasted the New Deal, survived the war, and perhaps found its fullest expression in the spirit of indulgent affection with which intellectuals often supported Harry Truman and his administration.

Hardly had Truman left the White House, however, when the Populist identification fell into disgrace and intellectuals began to repudiate the heritage. "Populist" suddenly became a term of opprobrium, in some circles a pejorative epithet. This resulted from no transfer of affection to Truman's successor, for there was very little of that among intellectuals. It resulted instead from the shock of the encounter with McCarthyism. Liberals and intellectuals bore the brunt of the degrading McCarthyite assault upon standards of decency. They were rightly alarmed and felt themselves betrayed. Something had gone badly wrong. They were the victims of a perversion of the democracy they cherished, a seamy and sinister side of democracy to which they now guiltily realized they had all along tended to turn a blind or indulgent eye. Stung by consciousness of their own negligence or naïveté, they reacted with a healthy impulse to make up for lost time and to confront their problem boldly with all the critical resources at their command. The consequence has been a formidable and often valuable corpus of social criticism.

Not one of the critics, not even the most conservative, is prepared to repudiate democracy. There is general agreement that the fault lay in some abuse or perversion of democracy, and was not inherent in democracy itself. All the critics are aware that these abuses and perversions had historic antecedents and had appeared in various guises and with disturbing frequency in national history. These unhappy tendencies are variously described as "mobism," "direct democracy," or "plebiscitarianism," but there is a surprising and apparently spontaneous consensus of preference for "Populism." Although the word is usually capitalized, most of the critics do not limit its reference to the political party that gave currency to the term. While there is general agreement that the essential characteristics designated by the term are best illustrated by an agrarian movement in the last decade of the nineteenth century, some of the critics take the liberty of applying it to movements as early as the Jacksonians, or earlier, and to twentieth-century phenomena as well.

The reasons for this convergence from several angles upon "Populism" as the appropriate designation for an abhorred abuse are not all clear. A few, however, suggest themselves. Populism is generally thought of as an entirely Western affair, Wisconsin as a seedbed of the movement, and Old Bob La Follette as a foremost exponent. None of these assumptions is historically warranted, but it is true that Senator McCarthy came from Wisconsin, that much of his support came from the Middle West, and that there are some similarities between the two movements. The impression of similarity has been enhanced by the historical echo of their own alarm that modern intellectuals have caught in the rather hysterical fright with which Eastern conservatives reacted to Populism in the nineties.

This essay is not concerned with the validity of recent analysis of the "radical right" and its fascistic manifestations in America. It is concerned only with the tendency to identify Populism with these movements and with the implied rejection of the Populist tradition. It is admittedly very difficult, without risk of misrepresentation and injustice, to generalize about the way in which numerous critics have employed the Populist identification. They differ widely in the meaning they attribute to the term and the importance they attach to the identification. Among the critics are sociologists, political scientists, poets and journalists, as well as historians, and there is naturally a diversity in the degree of historical awareness and competence they command. Among points of view represented are the New Conservative, the New Liberal, the liberal-progressive, the Jewish, the Anglophile, and the urban, with some overlapping. There are no conscious spokesmen of the West or the South, but some are more-or-less conscious representatives of the urban East. Every effort will be made not to attribute to one the views of another.*

Certain concessions are due at the outset. Any fair-minded historian will acknowledge the validity of some of the points scored by the new critics against the Populist tradition and its defenses.

---

* Daniel Bell (ed.), *The New American Right* (Criterion, 1955), especially essays by Richard Hofstadter, Peter Viereck, Talcott Parsons and Seymour Martin Lipset; Edward A. Shils, *The Torment of Secrecy* (Free Press, 1956) and "The Intellectuals and the Powers: Some Perspectives for Comparative Analysis," in *Comparative Studies in Society and History* I (October, 1958); Peter Viereck, *The Unadjusted Man* (Beacon, 1956); Oscar Handlin, *Race and Nationality in American Life* (Atlantic-Little, Brown, 1957), and "American Views of the Jews at the Opening of the Twentieth Century," *Publications of the American Jewish Historical Society*, no. 40 (June, 1951); Richard Hofstadter, *The Age of Reform* (Knopf, 1955); Victor C. Ferkiss, "Ezra Pound and American Fascism," *Journal of Politics*, XVII (1955); Max Lerner, *America as a Civilization* (Simon & Schuster, 1958).

It is undoubtedly true that liberal intellectuals have in the past constructed a flattering image of Populism. They have permitted their sympathy with oppressed groups to blind them to the delusions, myths and foibles of the people with whom they sympathized. Sharing certain political and economic doctrines and certain indignations with the Populists, they have attributed to them other values, tastes and principles that the Populists did not acually profess. It was understandably distasteful to dwell upon the irrational or retrograde traits of people who deserved one's sympathy and shared some of one's views. For undertaking this neglected and distasteful task in the spirit of civility and forebearance which, for example, Richard Hofstadter has shown, some of the new critics deserve much credit. All of them concede some measure of value in the Populist heritage, although none so handsomely as Hofstadter, who assumes that Populism and Progressivism are strongly enough established in our tradition to withstand criticism. Others are prone to make their concessions more perfunctory and to hasten on with the job of heaping upon Populism, as upon a historical scapegoat, all the ills to which democracy is heir.

The danger is that under the concentrated impact of the new criticism the risk is incurred not only of blurring a historical image but of swapping an old stereotype for a new one. The old one sometimes approached the formulation that Populism is the root of all good in democracy, while the new one sometimes suggests that Populism is the root of all evil. Uncritical repetition and occasional exaggeration of the strictures of some of the critics threaten to result in establishing a new maxim in American political thought: *Radix malorum est Populismus.*

Few of the critics engaged in the reassessment of Populism and the analysis of the New American Right would perhaps go quite so far as Peter Viereck, when he writes, "Beneath the sane economic demands of the Populists of 1880–1900 seethed a mania of xenophobia, Jew-baiting, intellectual-baiting, and thought-controlling lynch-spirit." Yet this far from exhausts the list of unhappy or repulsive aberrations of the American spirit that have been attributed to Populism. Other aberrations are not pictured as a "seething mania" by any one critic, but by one or another the Populists are charged with some degree of responsibility for Anglophobia, Negrophobia, isolationism, imperialism, jingoism, paranoidal conspiracy-hunting, anti-Constitutionalism, anti-intellectualism, and the assault upon the right of privacy, among others. The Populist virus is seen as no respecter of the barriers of time or nationality. According to Edward A. Shils, "populism has many faces. Nazi dictatorship had markedly populistic features. . . . Bolshevism has a strand of populism in it too. . . ." And there was among fellow travelers a "populistic predisposition to Stalinism." On the domestic scene the strand of populistic tradition "is so powerful that it influences reactionaries like McCarthy and left-wing radicals and great upperclass personalities like Franklin Roosevelt." And according to Viereck, populistic attitudes once "underlay Robespierre's Committee of Public Safety" and later "our neo-Populist Committee on un-American Activities."

Among certain of the critics there is no hesitancy in finding a direct continuity between the nineteenth-century Populists and twentieth-century American fascism and McCarthyism. Victor C. Ferkiss states flatly that "American fascism has its roots in American populism. It pursued the same ends and even used many of the same slogans. Both despaired of achieving a just society under the joined banners of liberalism and capitalism." His assertion supports Viereck's suggestion that "Since the same impulses and resentments inspire the old Populism and the new nationalist right, let us adopt 'neo-Populism' as the proper term for the latter group." Talcott Parsons believes that "The elements of continuity between Western agrarian populism and McCarthyism are not by any means purely fortuitous," and Edward Shils thinks the two are connected by "a straight line." It remained for Viereck to fill in the gap: "The missing link between the Populism of 1880–1900 and the neo-Populism of today—the missing link between Ignatius Donnelly and the McCarthy movement—was Father Charles Coughlin."

There is a strong tendency among the critics not only to identify Populism and the New Radical Right, but to identify both with certain regions, the West and South, and particularly the Middle West. "The areas which produced the populism of the end of the nineteenth century and the early twentieth century have continued to produce them," writes Shils. Viereck puts it somewhat more colorfully: "The Bible-belt of Fundamentalism in religion mostly overlapped with the farm-belt of the Populist, Greenback, and other free-silver parties in politics. Both belts were anti-intellectual, anti-aristocratic, anti-capitalist." Talcott Parsons and Ferkiss likewise stress the regional identity of Populist-Radical Right ideology, and Viereck supplies an interesting illustration: "Out of the western Populist movement came such apostles of thought-control and racist bigotry as Tom Watson. . . ."

If so many undesirable traits are conveniently concentrated along geographical lines, it might serve a useful purpose to straighten out the political geography of Populism a bit. In the first place, as Hofstadter and other historians of the movement have noted, Populism had negligible appeal in the Middle Western states, and so did the quasi-Populism of William Jennings Bryan. Wisconsin, Minnesota, Iowa, Illinois and states east of them went down the line for McKinley, Hanna, gold and the Old Conservatism (and so did Old Bob La Follette). Only in the plains states of the Dakotas, Nebraska and Kansas were there strong Populist leanings, and only they and the mountain states went for Bryan in 1896. At the crest of the Populist wave in 1894 only Nebraska polled a Populist vote comparable in strength to that run up in Alabama, Georgia and North Carolina.

For the dubious distinction of being the leading Populist section, the South is in fact a strong contender; and if the test is merely quasi-Populism, the pre-eminence of the former Confederacy is unchallengeable. It was easily the most solidly Bryan section of the country, and its dogged loyalty far outlasted that of the Nebraskan's native state. But a more important test was third-party Populism, the genuine article. The remarkable strength the Populists manifested in the Lower South was gained against far more formidable obstacles than any ever encountered in the West. For there they daily faced the implacable dogmas of racism, white solidarity, white supremacy and the bloody shirt. There was indeed plenty of "thought control and racist bigotry and lynch-spirit," but the Populists were far more often the victims than the perpetrators. They had to contend regularly with foreclosure of mortgages, discharge from jobs, eviction as tenants, exclusion from church, withholding of credit, boycott, social ostracism and the endlessly reiterated charge of racial disloyalty and sectional disloyalty. Suspicion of loyalty was in fact *the* major psychological problem of the Southern Populists, as much as perhaps as the problem of loyalty faced by radicals of today. They contended also against cynical use of fraud comparable with any used against Reconstruction, methods that included stuffed ballot boxes, packed courts, stacked registration and election boards, and open bribery. They saw election after election stolen from them and heard their opponents boast of the theft. They were victims of mobs and lynchers. Some fifteen Negroes and several white men were killed in the Georgia Populist campaign of 1892, and it was rare that a major election in the Lower South came off without casualties.

Having waged their revolt at such great cost, the Southern Populists were far less willing to compromise their principles than were their Western brethren. It was the Western Populists who planned and led the movement to sell out the party to the silverites, and the Southern Populists who fought and resisted the drift to quasi-Populism. The Southerners were consistently more radical, more insistent upon their economic reforms, and more stubbornly unwilling to lose their party identity in the watered-down populism of Bryan than were Western Populists.

There is some lack of understanding about *who* the Southern Populists were for and against, as well as *what* they were for and against. Edward Shils writes that the "economic and political feebleness and pretensions to breeding and culture" of the "older aristocratic ruling class" in the South provided "a fertile ground for populistic denunciation of the upper classes." Actually the Southern Populists directed their rebellion against the newer ruling class, the industrialists and businessmen of the New South instead of the old planters. A few of the quasi-Populists like Ben Tillman did divert resentment to aristocrats like Wade Hampton. But the South was still a more deferential society than the rest of the country, and the Populists were as ready as the railroads and insurance companies to borrow the prestige and name of a great family. The names of the Populist officials in Virginia sounded like a roll call of colonial assemblies or Revolutionary founding fathers: Page, Cocke, Harrison, Beverley, Ruffin. There were none more aristocratic in the Old Dominion. General Robert E.

Lee, after the surrender at Appomattox, retired to the ancestral home of Edmund Randolph Cocke after his labors. His host was later Populist candidate for governor of the state. As the editor of their leading paper, the allegedly Anglophobic Populists of Virginia chose Charles H. Pierson, an ordained Anglican priest, English by birth, Cambridge graduate and theological student of Oxford. To be sure, the Populist leaders of Virginia were not typical of the movement in the South. But neither were Jefferson, Madison, Monroe and John Taylor typical of *their* movement in the South: there were never enough aristocrats to go around. Some states had to make do with cruder customers as leaders in both Jeffersonian and Populist movements, and in the states to the west there doubtless was less habitual dependence on aristocrats even if they had been more readily available.

In their analysis of the radical right of modern America, the new critics have made use of the concept of "status resentment" as the political motivation of their subjects. They distinguish between "class politics," which has to do with the correction of economic deprivations, and "status politics," which has no definite solutions and no clear-cut legislative program but responds to irrational appeals and vents aggression and resentment for status insecurity upon scrapegoats—usually ethnic minorities. Seymour Martin Lipset, who appears at times to include Populism in the category, has outlined the conditions typical of periods when status politics become ascendant. These are, he writes, "periods of prosperity, especially when full employment is accompanied by inflation, and when many individuals are able to improve their economic position." But the conditions under which Populism rose were exactly the opposite: severe depression, critical unemployment and crippling currency contraction, when few were able to improve their economic position—and certainly not farmers in cash-crop staple agriculture.

The Populists may have been bitten by status anxieties, but if so they were certainly not bred on upward social mobility, and probably few by downward mobility either—for the simple reason that there was not much further downward for most Populists to go, and had not been for some time. Populism was hardly "status politics," and I should hesitate to call it "class politics." It was more nearly "interest politics," and more specifically "agri-

cultural interest politics." Whatever concern the farmers might
have had for their status was overwhelmed by desperate and immediate economic anxieties. Not only their anxieties but their
proposed solutions and remedies were economic. While their legislative program may have been often naïve and inadequate, it
was almost obsessively economic and, as political platforms go,
little more irrational than the run of the mill.

Yet one of the most serious charges leveled against the Populists in the reassessment by the new critics is an addition to just
the sort of irrational obsession that is typical of status politics.
This is the charge of anti-Semitism. It has been documented
most fully by Richard Hofstadter and Oscar Handlin and advanced less critically by others. The prejudice is attributed to
characteristic Populist traits—rural provinciality, and ominous
credulity and an obsessive fascination with conspiracy. Baffled
by the complexities of monetary and banking problems, Populist
ideologues simplified them into a rural melodrama with Jewish
international bankers as the principal villains. Numerous writings
of Western Populists are cited that illustrate the tendency to use
Jewish financiers and their race as scapegoats for agrarian resentment. Hofstadter points out that Populist anti-Semitism was
entirely verbal and rhetorical and cautions that it can easily be
misconstrued and exaggerated. Nevertheless, he is of the opinion
"that the Greenback-Populist tradition activated most of what
we have of modern popular anti-Semitism in the United States."

In the voluminous literature of the nineties on currency and
monetary problems—problems that were much more stressed by
silverites and quasi-Populists than by radical Populists—three
symbols were repetitively used for the plutocratic adversary. One
was institutional, Wall Street; and two were ethnic, the British
and Jewish bankers. Wall Street was by far the most popular and
has remained so ever since among politicians of agrarian and Populistic tradition. Populist agitators used the ethnic symbols more
or less indiscriminately, British along with Jewish, although some
of them bore down with peculiar viciousness on the Semitic symbol. As the new critics have pointed out, certain Eastern intellectuals of the patrician sort, such as Henry and Brooks Adams and
Henry Cabot Lodge, shared the Populist suspicion and disdain of
the plutocarcy and likewise shared their rhetorical anti-Semitism.

John Higham has called attention to a third anti-Semitic group of
the nineties, the poorer classes in urban centers. Their prejudice
cannot be described as merely verbal and rhetorical. Populists were
not responsible for a protest signed by fourteen Jewish societies
in 1899 that "No Jew can go on the street without exposing himself to the danger of being pitilessly beaten." That was in Brook-
lyn. And the mob of 1902 that injured some two hundred people, mostly Jewish, went into action in Lower East Side New
York.

Populist anti-Semitism is not to be excused on the ground that
it was verbal, nor dismissed because the prejudice received more
violent expression in urban quarters. But all would admit that
the charge of anti-Semitism has taken on an infinitely more ominous and hideous significance since the Nazi genocide furnaces
than it ever had before, at least in Anglo-American society. The
Populists' use of the Shylock symbol was not wholly innocent,
but they used it as a folk stereotype, and little had happened in
the Anglo-Saxon community between the time of Shakespeare
and that of the Populists that burdened the latter with additional
guilt in repeating the stereotype.

The South, again, was a special instance. Much had happened
there to enhance the guilt of racist propaganda and to exacerbate
racism. But anti-Semitism was not the trouble, and to stress it in
connection with the South of the nineties would be comparable
to stressing anti-Negro feeling in the Arab states of the Middle
East today. Racism there was, in alarming quantity, but it was
directed against another race and it was not merely rhetorical. The Negro suffered for more discrimination and violence
than the Jew did in that era or later. Moreover, there was little
in the Southern tradition to restrain the political exploitation
of anti-Negro prejudice and much more to encourage its use
than there was in the American tradition with respect to anti-
Semitism. Racism was exploited in the South with fantastic re-
finements and revolting excesses in the Populist period. Modern
students of the dynamics of race prejudice, such as Bruno Bettel-
heim and Morris Janowitz, find similarities between anti-Negro
feelings and anti-Semitism and in the psychological traits of
those to whom both appeal. First in the list of those traits under
both anti-Negro attitudes and anti-Semitism is "the feeling of

deprivation," and another lower in the list but common to both, is "economic apprehensions." The Southern Populists would seem to have constituted the perfect market for Negrophobia.

But perhaps the most remarkable aspect of the whole Populist movement was the resistance its leaders in the South put up against racism and racist propaganda and the determined effort they made against incredible odds to win back political rights for the Negroes, to defend those rights against brutal aggression, and to create among their normally anti-Negro following, even temporarily, a spirit of tolerance in which the two races of the South could work together in one party for the achievement of common ends. These efforts included not only the defense of the Negro's right to vote, but also his right to hold office, serve on juries, receive justice in the courts and defense against lynchers. The Populists failed, and some of them turned bitterly against the Negro as the cause of their failure. But in the efforts they made for racial justice and political rights they went further toward extending the Negro political fellowship, recognition and equality than any native white political movement has ever gone before or since in the South. This record is of greater historical significance and deserves more emphasis and attention than any anti-Semitic tendencies the movement manifested in that region or any other. If resistance to racism is the test of acceptability for a place in the American political heritage, Populism would seem to deserve more indulgence at the hands of its critics than it has recently enjoyed.

Two other aspects of the identification between the old Populism and the new radical right require critical modification. Talcott Parsons, Max Lerner and Victor Ferkiss, among others, find that the old regional strongholds of Populism tended to become the strongholds of isolationism in the period between the two world wars and believe there is more than fortuitous connection between a regional proneness to Populism and isolationism. These and other critics believe also that they discern a logical connection between a regional addiction to Populism in the old days and to McCarthyism in recent times.

In both of these hypotheses the critics have neglected to take into account the experience of the South and mistakenly assumed a strong Populist heritage in the Middle West. One of the strongest centers of Populism, if not the strongest, the South in the foreign policy crisis before the Second World War was the least isolationist and the most internationalist and interventionist part of the country. And after the war, according to Nathan Glazer and Seymour Lipset, who base their statement on opinion poll studies, "the South was the most anti-McCarthy section of the country." It is perfectly possible that in rejecting isolationism and McCarthyism the South was "right" for the "wrong" reasons, traditional and historical reasons. V. O. Key has suggested that among the reasons for its position on foreign policy were centuries of dependence on world trade, the absence of any concentration of Irish or Germanic population, and the predominantly British origin of the white population. Any adequate explanation of the South's rejection of McCarthy would be complex, but part of it might be the region's peculiarly rich historical experience with its own assortment of demagogues—Populistic and other varieties—and the consequent acquirement of some degree of sophistication and some minimal standards of decency in the arts of demagoguery. No one has attempted to explain the South's anti-isolationism and anti-McCarthyism by reference to its Populist heritage—and certainly no such explanation is advanced here.

To do justice to the new critique of Populism it should be acknowledged that much of its bill of indictment is justified. It is true that the Populists were a provincial lot and that much of their thinking was provincial. It is true that they took refuge in the agrarian myth, that they denied the commercial character of agricultural enterprise and sometimes dreamed of a Golden Age. In their economic thought they overemphasized the importance of money and oversimplified the nature of their problems by claiming a harmony of interest between farmer and labor, by dividing the world into "producers" and "nonproducers," by reducing all conflict to "just two sides," and by thinking that too many ills and too many remedies of the world were purely legislative. Undoubtedly many of them were fascinated with the notion of conspiracy and advanced conspiratorial theories of history, and some of them were given to apocalyptic premonitions of direful portent.

To place these characteristics in perspective, however, one

should inquire how many of them are peculiar to the Populists and how many are shared by the classes or groups or regions or by the periods to which the Populists belong. The great majority of Populists were provincial, ill-educated and rural, but so were the great majority of Americans in the nineties, Republicans and Democrats as well. They were heirs to all the superstition, folklore and prejudice that is the heritage of the ill-informed. The Populists utilized and institutionalized some of this, but so did their opponents. There were a good many conspiratorial theories and economic nostrums and oversimplifications adrift in the latter part of the nineteenth century, and the Populists had no monopoly of them. They did overemphasize the importance of money, but scarcely more so than did their opponents, the Gold Bugs. The preoccupation with monetary reforms and remedies was a characteristic of the period rather than a peculiarity of the Populists. The genuine Populist, moreover, was more concerned with the "primacy of credit" than with the "primacy of money," and his insistence that the federal government was the only agency powerful enough to provide a solution for the agricultural credit problem proved to be sound. And so did his contention that the banking system was stacked against his interest and that reform in this field was overdue.

The Populist doctrine of a harmony of interest between farmer and labor, between workers and small businessmen, and the alignment of these "producers" against the parasitic "nonproducers," is not without precedent in our political history. Any party that aspires to gain power in America must strive for a coalition of conflicting interest groups. The Populist effort was no more irrational in this respect than was the Whig coalition and many others, including the New Deal coalition.

The political crises of the nineties evoked hysterical responses and apocalyptic delusions in more than one quarter. The excesses of the leaders of a protest movement of provincial, unlettered and angry farmers are actually more excusable and understandable than the rather similar responses of the spokesmen of the educated, successful and privileged classes of the urban East. There would seem to be less excuse for hysteria and conspiratorial obsessions among the latter. One thinks of the *Nation* describing the Sherman Silver Purchase Act as a "socialistic contrivance of gigantic proportions," or of Police Commissioner Theodore Roosevelt declaring in "the greatest soberness" that the Populists were "plotting a social revolution and the subversion of the American Republic" and proposing to make an example of twelve of their leaders by "shooting them dead" against a wall. Or there was Joseph H. Choate before the Supreme Court pronouncing the income tax "the beginnings of socialism and communism" and "the destruction of the Constitution itself." For violence of rhetoric *Harper's Weekly*, the New York *Tribune* and the Springfield *Republican* could hold their own with the wool-hat press in the campaign of 1896. Hysteria was not confined to mugwump intellectuals with status problems. Mark Hanna told an assembly of his wealthy friends at the Union League Club they were acting like "a lot of scared hens."

Anarchism was almost as much a conspiracy symbol for conservatives as Wall Street was for the Populists, and conservatives responded to any waving of the symbol even more irrationally, for there was less reality in the menace of anarchism for capitalism. John Hay had a vituperative address called "The Platform of Anarchy" that he used in the campaign of 1896. The Springfield *Republican* called Bryan "the exaltation of anarchy"; Dr. Lyman Abbott labeled Bryanites "the anarchists of the Northwest," and Dr. Charles H. Parkhurst was excited about the menace of "anarchism" in the Democratic platform. It was the Populist sympathizer Governor John Peter Altgeld of Illinois who pardoned the three anarchists of Haymarket, victims of conservative hysteria, and partly corrected the gross miscarriage of justice that had resulted in the hanging of four others. The New York *Times* promptly denounced Governor Altgeld as a secret anarchist himself, and Theodore Roosevelt said that Altgeld would conspire to inaugurate "a red government of lawlessness and dishonesty as fantastic and vicious as the Paris Commune." There was more than a touch of conspiratorial ideology in the desperate conservative reaction to the agrarian revolt. An intensive study of the nineties can hardly fail to leave the impression that this decade had rather more than its share of zaniness and crankiness, and that these qualities were manifested in the higher and middling as well as the lower orders of American society.

Venturing beyond the 1890's and speaking of populists with a small p, some of the new critics would suggest that popular protest movements of the populistic style throughout our history have suffered from a peculiar addiction to scares, scapegoats and conspiratorial notions. It is true that such movements tend to attract the less sophisticated, the people who are likely to succumb to cranks and the appeal of their menaces and conspiratorial obsessions. But before one accepts this as a populistic or radical peculiarity, one should recall that the Jacobin Scare of the 1790's was a Federalist crusade and that the populistic elements of that era were its victims and not its perpetrators. One should remember also that A. Mitchell Palmer and the super-patriots who staged the Great Red Scare of 1919–1920 were not populistic in their outlook. One of the most successful conspiratorial theories of history in American politics was the Great Slave Conspiracy notion advanced by the abolitionists and later incorporated in the Republican Party credo for several decades.

Richard Hofstadter has put his finger on a neglected tendency of some Populists and Progressives as well, the tendency he calls "deconversion from reform to reaction," the tendency to turn cranky, illiberal and sour. This happened with disturbing frequency among leaders as well as followers of Populism. Perhaps the classic example is the Georgia Populist Tom Watson, twice his party's candidate for President and once for Vice-President. When Watson soured he went the whole way. By no means all of the Populist leaders turned sour, but there are several other valid instances. Even more disturbing is the same tendency to turn sour among the old Populist rank and file, to take off after race phobias, religious hatreds and witch hunts. The reasons for this retrograde tendency among reformers to embrace the forces they have spent years in fighting have not been sufficiently investigated. It may be that in some instances the reform movement appeals to personalities with unstable psychological traits. In the case of the Populists, however, it would seem that a very large part of the explanation lies in embittered frustration—repeated and tormenting frustration of both the leaders and the led.

Whatever the explanation, it cannot be denied that some of the offshoots of Populism are less than lovely to contemplate and rather painful to recall. Misshapen and sometimes hideous, they are caricatures of the Populist ideal, although their kinship with the genuine article is undeniable. No one in his right mind can glory in their memory, and it would at times be a welcome relief to renounce the whole Populist heritage in order to be rid of the repulsive aftermath. Repudiation of the Populist tradition presents the liberal-minded Southerner in particular with a temptation of no inconsiderable appeal, for it would unburden him of a number of embarrassing associations.

In his study of populist traits in American society, Edward Shils has some perceptive observations on the difficult relations between politicians and intellectuals. He adds a rather wistful footnote: "How painful the American situation looked to our intellectuals when they thought of Great Britain. There the cream of the graduates of the two ancient universities entered the civil service by examinations which were delightfully archaic and which had no trace of spoils patronage about them. . . . Politics, radical politics, conducted in a seemly fashion by the learned and reflective was wonderful. It was an ideal condition which was regretfully recognized as impossible to reproduce in the United States." He himself points out many of the reasons why this is possible in Britain, the most dignified member of the parliamentary fraternity: respect for "betters," mutual trust within the ruling classes, deferential attitudes of working class and middle class, the aura of aristocracy and monarchy that still suffuses the institutions of a government no longer aristocratic, the retention of the status and the symbols of hierarchy despite economic leveling. No wonder that from some points of view, "the British system seemed an intellectual's paradise."

America has it worse—or at least different. The deferential attitude lingers only in the South, and there mainly as a quaint gesture of habit. Respect for "betters" is un-American. Glaring publicity replaces mutual trust as the *modus vivendi* among the political elite. No aura of aristocratic decorum and hierarchal sanctity surrounds our governmental institutions, even the most august of them. Neither Supreme Court nor State Department nor Army is immune from popular assault and the rude hand of suspicion. The sense of institutional identity is weak, and so are institutional loyalties. Avenues between the seats of learning and the seats of power are often blocked by mistrust and mutual embarrassment.

America has no reason to expect that it could bring off a social revolution without a breach of decorum or the public peace, nor that the revolutionary party would eventually be led by a graduate of exclusive Winchester and Oxford. American politics are not ordinarily "conducted in a seemly fashion by the learned and reflective." Such success as we have enjoyed in this respect —the instances of the Sage of Monticello and the aristocrat of Hyde Park come to mind—have to be accounted for by a large element of luck. Close investigation of popular upheavals of protest and reform in the political history of the United States has increasingly revealed of late that they have all had their seamy side and their share of the irrational, the zany and the retrograde. A few of the more successful movements have borrowed historical reputability from the memory of the worthies who led them, but others have not been so fortunate either in their leaders or their historians.

One must expect and even hope that there will be future upheavals to shock the seats of power and privilege and furnish the periodic therapy that seems necessary to the health of our democracy. But one cannot expect them to be any more decorous or seemly or rational than their predecessors. One can reasonably hope, however, that they will not all fall under the sway of the Huey Longs and Father Coughlins who will be ready to take charge. Nor need they if the tradition is maintained which enabled a Henry George to place himself in the vanguard of the antimonopoly movement in his day, which encouraged a Henry Demarest Lloyd to labor valiantly to shape the course of Populism, or which prompted an Upton Sinclair to try to make sense of a rag-tag-and-bob-tail aberration in California.

For the tradition to endure, for the way to remain open, however, the intellectual must not be alienated from the sources of revolt. It was one of the glories of the New Deal that it won the support of the intellectual and one of the tragedies of Populism that it did not. The intellectual must resist the impulse to identify all the irrational and evil forces he detests with such movements because some of them, or the aftermath or epigone of some of them, have proved so utterly repulsive. He will learn all he can from the new criticism about the irrational and illiberal side of Populism and other reform movements, but he cannot afford to repudiate the heritage.

22   FROM   *Oscar Handlin*
              Reconsidering the Populists

The treatment of American populism in the last three decades reveals the extent to which history is still far from being a cumulative science which steadily refines the understanding of the past through successive scholarly discoveries. If any figure of speech is appropriate, it is that of the treadmill, with regressions and standstills as frequent as advances. A good deal has been written on the subject; progress, alas, has been slight.

Most of the difficulty arises from a pervasive Manichaeanism which interprets history as the battlefield between the forces of light and those of darkness—the good guys against the bad guys. In this drama, which can be pushed indefinitely back into the past, the honest, the altruistic and the patriotic constantly battle the greedy and the corrupt in defense of national virtue. This formulation has the virtue of simplicity and of clear identification of heroes and villains; its drawback lies in its slight correspondence with actuality.

The heated emotions the populist movement aroused among contemporaries have extended into the historical literature. Although the political issues of the 1890's have faded, the heat they engendered persists.

Well before the New Deal, the populists had captured the sympathy of the American historian.[1] Their agrarianism, their advocacy of the cause of the oppressed poor, their situation as game independents facing up to the entrenched and powerful, and their spontaneous grassroots origins gave them a heroic cast in the accounts of writers influenced by progressive ideas. Charles A. Beard summarized the prevailing attitude in *The Rise of American Civilization:* the populists were the culmination of the long history of agrarian protest and an integral element in the

SOURCE. *Agricultural History, 39*:68–74 (April 1965) .

[1] Van Woodward is in error in the chronology of the shifts in attitude toward the populists ("The Populist Heritage and the Intellectual," *American Scholar*, XXIX [1959–60], 55 ff.) .

struggle for American democracy. A more definitive academic statement of this position came from John Hicks in 1931. "Thanks to this triumph of Populist principles, one may almost say that, in so far as political devices can insure it, the people now rule. . . . the acts of government have come to reflect fairly clearly the will of the people."[2]

Such firm commitments have impeded every effort to interpret the character of the movement. Each bit of fresh evidence must meet the test of whether it casts the populists in the role of heroes or villains.[3] The result is a sterile round of argument that has added little to the understanding of the subject.

The extended—and pointless—discussion of whether the populists were anti-Semitic is a glaring illustration.

• • •

Apologia has obscured the genuine problems that the scholars now concerned with populism ought to address. Once we cease to worry about whether the populists were more farsighted and more virtuous than their contemporaries, we shall be able to confront the characteristics really distinctive of the movement.

Populism drew together many discontented elements in American society—farmers, laborers, social reformers and intellectuals. The very diversity of these groups renders futile the quest for a single ideology among them. Socialists and currency men, Bellamy Nationalists and Knights of Labor could unite in complaints against the contemporary capitalism and in resentment of the moral deficiencies of the new wealth but not in their vision of the future; significantly, Norman Pollack's book which seeks to define such an ideology deals with criticism of existing conditions, not with proposals for the future.[4]

[2] Charles A. and Mary R. Beard, *Rise of American Civilization* (New York, 1927) , II, 278 ff., 556 ff.; John D. Hicks, *Populist Revolt* (Minneapolis, 1931) , 422; John D. Hicks, "The Persistence of Populism," *Minnesota History*, XII (1931) , 3 ff.; John D. Hicks, "The Legacy of Populism in the Western Middle West," *Agricultural History*, XXIII (1949) , 225 ff.

[3] See the candid account in Walter T. K. Nugent, *The Tolerant Populists; Kansas Populism and Nativism* (Chicago, 1963) , 3 ff.

[4] Pollack, *Populist Response, passim;* Destler, "Agricultural Readjustment," *Agricultural History*, XXI, 115; Chester M. Destler, *American Radicalism 1865–1901* (New London, 1946) , 15 ff.

The process of voicing these grievances produced not a common set of ideas but a common style. Populist literature and oratory were laden with a rhetoric that appealed to the emotions. The audience was again and again summoned to "put on the bright armour of chivalry, ride forth to the rescue and smite the dungeon-door with the battle-axe of Lionheart." It therefore responded with "impulsiveness and emotionalism" to "anything from an economic doctrine to a political platform." What was true of Kansas was often true elsewhere; the election of 1890 could "hardly be diagnosed as a political campaign. It was a religious revival, a crusade, a pentecost of politics in which a tongue of flame sat upon every man, and each spake as the spirit gave him utterance." Senator Peffer once explained that "the Alliance is in a great measure taking the place of the churches"; he was correct in more ways than he knew. The style had recognizable antecedents as far back as the Great Awakening. And it was particularly compelling for those in the audience reared in the evangelical denominations.[5]

Sensational revelations of conspiracies which threatened the safety of the republic spiced the treatises on money or on the trusts. The authors and speakers were, of course, sincere; they believed absolutely in the gospel for which they sought converts. But the villainies they exposed did not become known through research or investigation; although some of the authors worked industriously to compile their data, in the end the conspiracy of the gold gamblers could only be inferred. But it had to be true because only thus could the discrepancy between events and faith be reconciled.[6]

The populist faith rested consistently upon one article: the competence of the common man—in whom a divine spark dwelt —to deal adequately with all the problems of life. Control by the mass of the people could solve every difficulty; the direct election of Senators, the primary, cooperatives and government owner-

[5] C. Vann Woodward, *Tom Watson, Agrarian Rebel* (New York, 1938), 41 138, 139; Hicks, *Populist Revolt*, 159.

[6] Norman Pollack is himself the victim of the same faulty logic, "Hofstadter on Populism," *Journ. So. Hist.*, XXVI, 493; Paul W. Glad, *Trumpet Soundeth; William Jennings Bryan and His Democracy, 1896–1912* (Lincoln, 1960), 49 ff.

ship were ultimately means of establishing that control. The voice of the people was the voice of God, even when it would, later, be sounded by a lynching mob.[7]

Most populists in the 1890's, however, learned to accept a distinction between the people as they were and as they should be, if for no other reason than because the majority voted incorrectly. "The very fool workingmen who are deprived of the means of subsistence will continue to vote for the perpetuation of the system which is constantly adding to the number of unemployed just as often as they get a chance," wrote the Topeka *Advocate*.[8]

The populist could only explain away the failure of the people to elect those who would best serve them by references to the corrupting power of conspiracy. And he had no doubt about who were the unregenerate and who the redeemed. "The government is the *people* and *we* are the people," announced Jerry Simpson. "Old man Peepul is on top. Aunt Sarah Jane is on top. We country folks are on top, and everybody is going to be happy." Virtue was safe on the farm, in danger in the urban setting. An agrarian promised land was the true home of the people; the apocalyptic visions of *Caesar's Column* exposed the alternative. Hence the hostility to the city, the resentment of its power to draw men from the farms and the mistrust of the strange populace that clustered there.[9]

The awareness that the people as they were were not the same as the people as they should have been also turned many populists against the institutions which deceived and misled as well as oppressed the common man. The business corporations, entrenched in Wall Street, were an obvious target; but the mistrust and hostility also spilled over to the hireling subsidized press, to the universities and occasionally to the churches which were all tools of the capitalists and unresponsive to the popular will.

One can neither accept these categorizations at face value as valid nor dismiss them offhand as aberrations. Properly viewed, the relationship of the populists to the social institutions of their

[7] Woodward, *Watson*, 445; R. C. Miller, "Background of Populism in Kansas," *Mississippi Valley Historical Review*, XI (1925), 469 ff.; Glad, *Trumpet Soundeth*, 34.

[8] Pollack, *Populist Response*, 62.

[9] Nugent, *Tolerant Populists*, 77; Woodward, *Watson*, 428.

complex society can shed light on both the movement and the objects of its attack.

The bitterness of the attack is comprehensible in terms of the lack of understanding on the part of the opponents of the populists. The scorn and sometimes the practical penalties visited upon the dissidents were enough to account for their resentment. But, in addition, significant changes in the institutional life of the nation made the position of those who did not fit uncomfortable. The elaboration of a formal network of organization and of communication, the articulation of particular elements into a highly structured system, professionalism, and the appearance of centralized controls were novel and tended to separate institutions from the life of the people. But these developments were not the products of any conscious plot, nor did they result in the dominance of the plutocrats. The populists sensed, but neither understood nor knew how to deal with, the growing distance between institutions and people.[10]

Their sweeping mistrust of the social and cultural authority being established in these years led to a total reliance upon common sense, unfettered by external discipline. The professors who were wrong on the question of silver were also in error on the authorship of *Hamlet*. The same kind of reasoning led Donnelly to his conclusions in both cases. The eccentricities referred to earlier were the direct product of the rejection of the standards sustained by the dominant institutions of the nation. More important, the populists could not perceive that within the respectable churches and universities and even within some of the great business corporations important forces were at work to give these institutions a new sense of social purpose.

The political party was the institution with which the populists were most involved—first in the effort to capture control from within, then in the formation of a separate organization, and finally in the fusion of 1896. Yet while the course of specific campaigns has been exhaustively studied, we still lack an adequate analysis of the political party as a functioning institution in this period. It is significant that Professor Hicks' judgment has been

[10] See, e.g., Simkins, *Tillman Movement*, 142 ff.; Ridge, *Donnelly*, 209, 264, 266, 370.

allowed to stand: "Populist propaganda in favor of independent voting did much to undermine the intense party loyalties that had followed in the wake of the Civil War." Indeed, Professor Pollack seems to assume that there was "a need for the formation of a third party" in the 1890's, although he also argues that there was not a need for a third party in 1896.[11]

Yet in these years the Democratic and Republican organizations took their modern form and the two-party system became firmly entrenched in American politics. The pressures that operated on the dominant parties also affected the populists and—in ways that have not yet been explored—markedly influenced both tactics and objectives. For instance, an examination of the situation in the national and state legislatures of the populists as politicians and not as reformers would throw light on the fluctuations in their support in 1894 and on the maneuvers that led to fusion two years later. Professor Hollingsworth has called attention to the failures of leadership in this period; we need also to know why so much more was demanded of it.[12]

Above all, we need to know far more than we do about the audience that responded to populist appeals. Their statements of grievances are eloquent evidence of the economic hardships these people suffered. But this is partial evidence only and must be filled out from other sources. Complaints of debt are frequent, for instance; but we have only fragmentary information on the process by which indebtedness snowballed. Ben Tillman traced his personal downfall back to 1881 when "the devil tempted me to buy a steam engine and other machinery, amounting to two thousand dollars, all on credit." The devil, in the guise of a city slicker, has remained a convenient scapegoat; one scholar has concluded that an "avalanche of credit" from the east tempted the western farmers "to extravagance, over-investment and speculation.[13]

Such explanations tell only half the story. Why did some agriculturalists and not others yield to temptation? Why was the in-

---

[11] Hicks, *Populist Revolt*, 409; Pollack, "Hofstadter on Populism," *Journ. So. Hist.*, XXVI, 487, 498.

[12] J. Rogers Hollingsworth, *The Whirligig of Politics* (Chicago, 1963), 235 ff.; Woodward, *Watson*, 278ff.; George H. Mayer, *The Republican Party* (New York, 1964), 243 ff.; Stanley L. Jones, *Presidential Election of 1896* (Madison, 1964), 77 ff., 246 ff.; Martin, *People's Party in Texas*, 141 ff.

[13] Simkins, *Tillman Movement*, 51; Hicks, *Populist Revolt*, 23.

cidence of debt high not in the poorest sections but in those in which land values rose rapidly? Studies like those of Allan Bogue reveal that interest rates were falling and that it was by no means a simple matter to attract eastern or European funds into the west. And the farmers were by no means reluctant victims. Tillman himself recalled that, "My motto was, 'It takes money to make money, and nothing risk nothing have'." Denunciations of the competitive system were readily heard in hard times, but a speculative, acquisitive atmosphere seemed characteristic of many regions just before they turned populist and just after they left the faith. It would help to know to what extent the farmers —and editors, lawyers, doctors and small-town businessmen— who became populists remained spiritually boomers and sooners.[14]

We catch only fragmentary glimpses, beneath the definable economic grievances, of more pervasive causes of discontent. However idyllic the images of the happy husbandman, the audience knew well enough the isolation of the Great Plains, the debilitating effects of illness, the devastation of climatic disasters. No change in railroad rate or cotton prices could transform these harsh conditions. The alliances, like the granges earlier, were social before they were political organizations; and perhaps the intensity of the emotions of their members owed something to the repressed consciousness that some among their problems were utterly beyond their control.[15]

In sum, what is called for is an effort to understand rather than to defend or attack the populists. They were neither saints nor sinners, but men responding to the changes that were remaking America in their time. They have to be comprehended in that light.

[14] Allan G. Bogue, *Money at Interest* (Ithaca, 1955), 83 ff., 263 ff., 269 ff.; Allan G. Bogue, *From Prairie to Corn Belt; Farming on the Illinois and Iowa Prairies in the Nineteenth Century* (Chicago, 1963), 177; Malin, "Mobility and History," *Agricultural History*, XVII, 181 ff.; Simkins, *Tillman Movement*, 51 ff.; Pollack, *Populist Response*, 18, 27; Nugent, *Tolerant Populists*, 54 ff.; Hofstadter, *Age of Reform*, 42, 58; Martin Ridge, *Ignatius Donnelly*, 17 ff.; Glad, *Trumpet Soundeth*, 46; Woodward, "Populist Heritage and the Intellectual," *American Scholar*, XXIX, 62.

[15] Hicks, *Populist Revolt*, 128 ff.; William D. Sheldon, *Populism in the Old Dominion* (Princeton, 1935), 40 ff.; Theodore Saloutos, *Farmer Movements in the South 1865–1933* (Berkeley, 1960), 69 ff.

PART FIVE

Bibliographical Note

General bibliographies dealing with American farmers and farming are Everett E. Edwards, *A Bibliography of the History of Agriculture in the United States* (Washington, D.C. 1930. Reprinted, Detroit, Mich. 1967) and John T. Schlebecker, *A Bibliography of Books and Pamphlets on the History of Agriculture in the United States . . . 1607–1967* (Santa Barbara, Calif. 1969). Dennis S. Nordin, *A Preliminary List of References for the History of the Granger Movement* (Davis, Calif., 1967) is reasonably complete on the subject to the date of publication. George H. Miller, *Railroads and the Granger Laws* (Madison, Wis. 1971) provides additional material in the bibliographic essay, pp. 263–286. Denton E. Morrison, ed., *Farmers' Organizations and Movements: Research Needs and a Bibliography of the United States and Canada* (Research Bulletin, Agricultural Experiment Station, East Langing, Mich.) offers an extensive list of books, articles and pamphlets on farmer organizations in the United States. Carl C. Taylor, *The Farmers' Movement, 1620–1920* (New York, N.Y. 1953) includes a wide ranging bibliography. Robert L. Tontz, "Membership of General Farmers' Organizations, United States, 1874–1860," *Agricultural History* 38:143–156 (July 1964) brings together useful figures showing the size of farmer organizations.

Liberty Hyde Bailey, ed., *Cyclopedia of American Agriculture: A Popular Survey of Agricultural Conditions, Practices and Ideals in the United States and Canada,* 4 volumes,

(New York, N.Y., 1909) provides a mass of information about farms and farming at the beginning of the 20th century. Fred A. Shannon, *The Farmer's Last Frontier: Agriculture 1860–1897* (New York, N.Y. 1945) is a general history of American agriculture from the time of the Civil War to the end of the century. Gilbert C. Fite, *The Farmers' Frontier 1865–1900* (New York, N.Y. 1966) concentrates on new land settlement during the period. Earl W. Hayter, *The Troubled Farmer 1850–1900: Rural Adjustment to Industrialization* (DeKalb, Ill., 1968) deals with many of the difficulties farmers, particularly those of the Middle West, encountered in adjusting to extensive change. Many of the technological changes in farming are dealt with by Leo Rogin, *The Introduction of Farm Machinery in its Relation to the Productivity of Labor in the Agriculture of the in the United States during the Nineteenth Century* (Berkeley, Calif. 1931) and Clarence H. Danhof, *Change in Agriculture: The Northern United States, 1820–1870* (Cambridge, Mass. 1969). Eric E. Lampard, *The Rise of the Dairy Industry in Wisconsin: A Study of Agricultural Change, 1820–1920* (Madison, Wis., 1963) and Stanley N. Murray, *The Valley Comes of Age: A History of Agriculture in the Valley of the Red River of the North, 1812–1920* (Fargo, N.D. 1967) provide detailed studies of agricultural change and accommodation from the time of first settlement to the early years of the 20th century.

Chester M. Destler, *American Radicalism, 1863–1901* (New London, Conn. 1964) and Richard Hofstader, *The Age of Reform* (New York, N.Y. 1955) deal with aspects of post Civil War reform movements. James Dabney McCabe, under the pseudonym, Edward Winslow Martin, *History of the Grange Movement* (Philadelphia, Pa. 1873. Reprinted, with introduction by Michael Hudson, New York, N.Y., 1969), Jonathan Periam, *The Groundswell: A History of the Origins, Aims, and Progress* of the Farmers' Movement. . . . (Cincinnati and Chicago, 1874) and Oliver H. Kelley, *Origins and Progress of the Order of the Patrons of Husbandry in the United States: A History from 1866–1873* (Philadelphia, Pa. 1875) are the most detailed contemporary accounts of the origins and purposes of the Granger Movement and the rise and spread of the Patrons of Husbandry. Solon J. Buck, *The Granger Movement: A Study of*

Agricultural Organization and its Political Economic, and Social Manifestations, 1870–1880 (Cambridge, Mass., 1913. Paperback edition, Nebraska, 1963) was the first extended, scholarly study of the Granger Movement. Studies such as Lee Benson, *Merchants, Farmers, and Railroads: Railroad Regulations and New York Politics, 1850–1887* (Cambridge, Mass., 1955), Gabrial Kolko, *Railroads and Regulation 1877–1916* (Princeton, N.J., 1965), and George H. Miller, *Railroads and the Granger Laws* (Madison, Wis., 1971) extend and modify Buck's work. Irwin Unger, *The Greenback Era* (Princeton, N. J., 1964) and Allen Weinstein, *Prelude to Populism: Origins of the Silver Issue* (New Haven, Conn., 1970) touch on farmer concern with monetary questions.

W. Scott Morgan, *History of the Wheel and Alliance and the Impending Revolution* (Fort Scott, Kansas, 1889. Reprint of the 1891 edition, New York, N.Y. 1968) and N. A. Dunning, *Farmers' Alliance History and Agricultural Digest* (Washington, D.C. 1891) offer contemporary, detailed accounts of the rise and purposes of the Alliance movement. John D. Hicks, *The Populists Revolt: A History of the Farmers' Alliance and the People's Party* (Minneapolis, Minn., 1931. Paperback, University of Nebraska, Lincoln, Neb., 1961) is the first thoroughgoing scholarly treatment of the Alliance-Populist movement. This has been followed by a number of other studies including Norman Pollack, *The Populist Response to Industrial America; Midwestern Populist Thought* (Cambridge, Mass., 1962) and *The Populist Mind* (New York, N.Y., 1967), Anna Rochester, *The Populist Movement in the U.S.* (New York, N.Y., 1943), and Theodore Saloutos, *Farmer Movements in the South, 1865–1933* (Berkeley, Calif., 1960). A large number of specialized or regional studies include O. Gene Clanton, *Kansas Populism: Ideas and Men* (Lawrence, Kans., 1969), Robert F. Durden, *The Climax of Populism: The Election of 1896* (Lexington, Ky., 1965) W. I. Hair, *Bourbonism and Agrarian Protest: Louisiana Politics, 1877–1900* (Baton Rouge, La., 1969) Sheldon Hackney, *Populism to Progressivism in Alabama* (Princeton, N.J., 1969) Frederick E. Haynes, *Third Party Movements Since the Civil War, with Special to Iowa; A Study in Social Politics* (Iowa City, Ia., 1916), Walter T. K. Nugent, *The Tolerant Populists: Kan-*

sas Populism and Nativism (Chicago, Ill., 1963), William Rogers, *The One-Gallused Rebellion: Agrarianism in Alabama, 1865–1896* (Baton Rouge, La., 1970), and R. C. Martin, *The People's Party in Texas: A Study in Third Party Politics* (Austin, Tex., 1933).

In addition to the biographical sketches of farmer and reform leaders to be found in the *Dictionary of American Biography*, several full-length biographies are particularly useful: Frederick E. Haynes, *James Baird Weaver* (Iowa City, Ia., 1919), Caroline A. Lloyd, *Henery Demarest Lloyd, 1847–1903* (New York, N.Y., 1912), Stuart Noblin, *Leonidas LaFayette Polk, Agrarian Crusader* (Chapel Hill, N.C., 1949), Martin Ridge, *Ignatius Donnelly, A Portrait of a Politician* (Chicago, Ill., 1962), Francis B. Simkins, *Pitchfork Ben Tillman, South Carolinian* (Baton Rouge, La., 1944), and C. Vann Woodward, *Tom Watson: Agrarian Rebel* (New York, N.Y., 1935).

Schlebecker's *Bibliography* includes a number of novels that deal with life on the American farm. Such works as Willa S. Cather, *My Antonia,* Hamlin Garland, *Son of the Middle Border,* Frank Norris, *The Octopus,* Ole E. Rolvaag, *Giants in the Earth,* Mari Sandoz, *Old Jules* (a fictionized biography), and T. S. Stribling, *The Store* provide some sharply drawn pictures of farm life after the Civil War.

Collections of documents dealing with farmer movements are Fred A. Shannon, ed., *American Farmers' Movements* (Princeton, N.J., 1957), Raymond J. Cunningham, ed., *The Populists in Historical Perspective* (New York, N.Y., 1968), Theodore Saloutos, ed., *Populism: Reaction or Reform?* (New York, N.Y. 1968) George B. Tindall, ed., *A Populist Reader: Selections from the Works of American Populist Leaders* (New York, N.Y., 1968) and Irwin Unger, ed., *Populism: Nostalgic or Progressive* (Berkeley, Calif. 1964).